With its clear prose and incisive analysis, *The Making of Low Carbon Economies* provides new insight into the "multiple ways in which climate change is being made economic". In this important book, Heather Lovell provides us with a new perspective from which to view the carbon economy, moving beyond the boundaries of carbon markets and individual policy instruments to examine how diverse sites, from new homes to old heating systems, forests to accounting practices, are becoming part of carbon economies. Her original approach places the materiality of carbon markets – the technologies, techniques, and artefacts through which they are assembled, framed and made commensurate – centre stage and demonstrates the complex and often fragile ways in which carbon comes to matter economically. *The Making of Low Carbon Economies* provides a genuinely new account of the carbon economy that will have important implications for all who work in the field of climate policy and politics.

Professor Harriet Bulkeley, *Department of Geography, University of Durham, UK*

With admirable clarity and precision, Lovell brings to life how responding to climate change is reorganising whole patterns of economic and social organisation. She shows how practices of measuring, counting, comparing and mobilising carbon are transforming markets, buildings, forests and accountancy, at the same time introducing us with a deft, light touch, to a wide range of complex theoretical tools needed to understand this transformation. This is a must-read for anyone who wants to understand the complexity produced by our emerging attempts to decarbonise society, and provides insights for the established researcher as well as the undergraduate student in a way few authors are able to achieve.

Professor Matthew Paterson, *School of Political Studies, University of Ottowa, Canada*

How does a polluting by-product of industrial societies become a commodity and what work does it do? In this clear-sighted book, Heather Lovell analyses the way "carbon" markets are constructed and how they perform economically. The messy realities of dealing with carbon take the reader far from assumptions in economic textbooks. In-depth empirical studies of the underlying details of weaving carbon into a range of activities provides some refreshing insights into the sociological, political and material workings of making carbon economic.

Dr Adrian Smith, *SPRU (Science and Technology Policy Research), University of Sussex, UK*

THE MAKING OF LOW CARBON ECONOMIES

The Making of Low Carbon Economies looks at how more than two decades of sustained effort at climate change mitigation has resulted in a variety of new ideas, institutions, regulations and ways of doing things: a period of active construction of low carbon economies. The book takes a fresh look at society's response to climate change by examining a diverse array of empirical sites where climate change has been incorporated into everyday lives – a process of stitching climate concerns into the discourse and practices of already existing economies, as well as creating new ones. *The Making of Low Carbon Economies* uses three interdisciplinary themes – heterogeneous networks, framing and commensuration – drawn variously from sociology, human geography, Foucauldian scholarship and Science and Technology Studies, to explore questions about how, where and why low carbon economies are being made.

Written by an acknowledged expert in the field, this is an original and important contribution to the scholarly debate on climate change. The book will be of interest to academics and higher-level students interested in climate change and the transition to a low carbon society.

Heather Lovell is a Reader in the Institute of Geography and the Lived Environment in the School of Geosciences at Edinburgh University, UK. Her research focuses on low energy housing, and the commodification of carbon. She has been funded by the ESRC, EPSRC and UK Nuffield Foundation, and is currently co-investigator on two UK Research Council projects examining domestic energy demand.

THE MAKING OF LOW CARBON ECONOMIES

Heather Lovell

Routledge
Taylor & Francis Group
LONDON AND NEW YORK

First published 2015
by Routledge
2 Park Square, Milton Park, Abingdon, Oxon OX14 4RN

and by Routledge
711 Third Avenue, New York, NY 10017

Routledge is an imprint of the Taylor & Francis Group, an informa business

British Library Cataloguing in Publication Data
A catalogue record for this book is available from the British Library

Library of Congress Cataloging in Publication Data
Lovell, Heather.
The making of low carbon economies / Heather Lovell.
pages cm
Includes bibliographical references and index.
1. Carbon dioxide mitigation–Economic aspects. 2. Carbon offsetting–Economic aspects. 3. Environmentalism–Economic aspects. 4. Sustainable development–Environmental aspects. 5. Climatic changes–Economic aspects. 6. Climatic changes–Social aspects. I. Title.
HC79.P55L69 2015
363.738'74–dc23
2014025145

ISBN: 978-0-415-72471-5 (hbk)
ISBN: 978-1-315-85704-6 (ebk)

Typeset in Perpetua
by Wearset Ltd, Boldon, Tyne and Wear

CONTENTS

	List of plates	viii
	List of tables	ix
	Preface	x
	Acknowledgements	xii
	List of abbreviations	xv
1	Introduction	1

PART I
Heterogeneous networks **13**

2	Making carbon offsets	27
3	Implementing residential district heating	52

PART II
Framing **71**

4	Low carbon housing in the UK	81
5	The frames of carbon accounting	104

PART III
Commensuration **125**

6	Measuring and standardising forest carbon for REDD+	133
7	The treatment of carbon in financial accounts	156
8	Summary and conclusions	174
	Index	180

PLATES

3.1 Installation of district heating pipework at Byford, early 2012 57
3.2 The heating control panel installed at Fordside 65
6.1 A backhoe digger excavating tree roots in order to measure the total amount of carbon stored in a sample of trees, Miombo woodlands, Mozambique 134

TABLES

2.1 Technologies used to produce carbon offsets in the Clean Development Mechanism 33
2.2 Further breakdown of selected Clean Development Mechanism project technologies by sub-type 34
3.1 Key general characteristics of Byford and Fordside 56
4.1 Belief system of the sustainable housing advocacy coalition 85
4.2 Principle founding actors of the UK sustainable housing advocacy coalition 86
4.3 The framing of sustainable housing as a solution to UK policy problems at the start of the twenty-first century 88
5.1 A working definition of carbon accounting 118
6.1 Examples of use of the Measurement, Reporting and Verification (MRV) storyline by diverse organisations 142
6.2 GOFC-GOLD author survey findings 146
6.3 Examples of forest carbon MRV standards 148
7.1 Extracts of the IASB 2011 Agenda Consultation response letters related to emissions trading 168

PREFACE

This book was originally going to be just about carbon markets – the relatively discrete markets explored in Chapter 2 (and some of their accounting and measurement implications, discussed in Chapters 6 and 7). I was fortunate enough to be awarded a three-year fellowship from the UK Nuffield Foundation to work on carbon markets (2008–2011), called "Fungible carbon". However, by the time my fellowship was drawing to a close, the world of carbon markets looked very different: there was no follow-on international climate agreement and the value of carbon (the neatly defined commodity of carbon traded within carbon markets) had plummeted. I remembered something the director of a leading carbon offset company had revealed in an interview back in 2007 – a point when he was already becoming disillusioned with carbon markets (having been active in the field since 1997, one of the founding figures). He said that next he would like to design and market a low carbon product. No complex carbon market rules, no process of establishing baselines or additionality – just a product that would do the work of reducing greenhouse gas emissions. Coincidentally, shortly after this, I stumbled upon a report by the UK Green Building Council about an aspect of the 2016 zero carbon homes policy called 'Allowable solutions' (see Chapter 4). In this report, I was surprised to find myself reading about offsets and additionality, as part of a proposal to allow new UK housing developments to be 'zero carbon' by generating credits from renewable energy technologies located elsewhere. These two things got me thinking about the blurred boundaries between carbon markets and low carbon economies, and the different ways in which 'carbon' – as a new type of commodity designed to mitigate climate change – has manifested. The seed of an idea for a book, covering several case studies of the making of diverse low carbon economies, was planted.

This book also has a practical origin, in that I have worked across a range of empirical areas and topics to do with climate mitigation since commencing my PhD in 2001. My PhD and postdoctoral fellowship focused on housing, energy and climate change (2001–2006). In 2007, I started a job at the UK Tyndall

Centre, working on carbon offsets and carbon markets. Getting to grips with this new subject area – complete with an overly abundant range of acronyms (CERs, CDM, ETS, AAU . . .) – encouraged me to think about applying for a grant on carbon markets, and in 2008 I was awarded a Nuffield Foundation New Career Development Fellowship ("Fungible carbon") at Edinburgh University. Keen to understand how and why the (carbon market) commodity of carbon had travelled, this grant enabled me to hang out with financial accountants, as well as forest remote sensing scientists. Running concurrently was a project I was involved in investigating district heating in the cities of Edinburgh and Glasgow ("Heat and the city" (2010–2014)). 'Just stick to one thing' was the advice that rang in my head, 'people don't like academics changing fields'. At busy moments, this rang true. But to me, it never felt like I was working across different fields, as there was a core body of theory I was using that spoke to all of my findings. I came to think more and more about the commonalities and synergies across the different sites and topics of climate mitigation that I have been fortunate enough to research. This book is the result.

ACKNOWLEDGEMENTS

Many individuals and organisations have been involved in this book. First, I wish to say thank you to those who have supervised and mentored me in my research since I commenced on an academic career in 2001: Professor Susan Owens for supervising my doctoral research at Cambridge University (2001–2005); Professor Susan Smith for her mentoring on a subsequent postdoctoral fellowship at Durham University (2005–2007); Professor Diana Liverman in her role as Director of the Tyndall Centre project, which I worked on as a postdoctoral researcher (2007); Professor Donald MacKenzie for his mentoring during our joint Nuffield Foundation New Career Development Fellowship on carbon markets and carbon accounting (2008–2011). I am extremely grateful to all of them for their inspiration, positive example, encouragement and patience, which, in all cases, went well beyond their formal role.

Second, I wish to thank those organisations who have funded the climate change research I have conducted, without whom this book would not have been possible. To the UK Economic and Social Research Council (ESRC) for my doctoral grant (R42200134546) and postdoctoral grant (PTA-026–27–0783) for research on UK low energy housing ("The governance of emerging socio-technical systems: the case of low energy housing in the UK") and also for the grant "Heat and the city" (RES-628–25–0052), a project led by Professor Webb at Edinburgh University, which I am a co-investigator on, and which provides the basis for Chapter 3. Particular thanks are due to the Nuffield Foundation for the New Career Development Fellowship they awarded me in 2008, "Fungible carbon" (NCF/35037).

Third, thanks to those who have read drafts of all or parts of this book, especially Dr Martin Pullinger, who read an earlier version of the whole book and provided very helpful comments; likewise, thanks go to Dr Francisco Ascui, whose strong attention to detail and tireless encouragement made a huge difference. Professor Harriet Bulkeley helped me enormously in refining the idea for this book on a train journey together in June 2012 – thank you. The Chisholm

House Energy Reading Group at Edinburgh University also deserve a mention here, for their insightful comments on drafts of the core conceptual material (Parts I, II and III) and Chapter 3 (especially David McCrone, Jan Webb, Ronan Bolton, Dave Hawkey and Mags Tingey). Also, thanks to my father, Dr Bryan Lovell, for his enthusiastic and speedy comments on Chapters 1, 2 and 8; and to Dr Ben Evar, for his insightful comments on several chapters. I am also extremely grateful to the team at Routledge for their help on this book, in particular Sarah Gilkes and Elizabeth Welsh. All mistakes that remain are, of course, my own.

I wish also to express gratitude to colleagues who have worked directly with me on previous versions of chapters of this book – either assisting with gathering empirical evidence and/or analysis and writing – including: Professor Jan Bebbington, Professor Carlos Larrinaga, Dr Thereza Sales de Aguiar and Xuan Sheng Ou Yong (Chapter 7); Professor Iain Woodhouse, Professor Martin Herold and Dr Brice Mora (Chapter 6); Professor Jan Webb, Professor David McCrone, Dr Dave Hawkey and Mags Tingey (Chapter 3); Professor Diana Liverman (Chapter 2); and Dr Francisco Ascui (Chapter 5).

Fourth, there are many other colleagues I would like to thank who I have worked with over the last fifteen years, too numerous to mention all by name here, but I particularly thank those who have made me laugh. Academic research is, of course, a serious endeavour, but I have found humour to help in all sorts of ways.

Fifth, practitioners and others outside of academia involved in this research are owed a special mention, for their generosity in sparing the time to be involved, be it through interviews, email exchanges, surveys or otherwise. Particular thanks is given (in no particular order) to: GOFC-GOLD, the Climate Disclosure Standards Board, the Association for Certified Chartered Accountants (ACCA), the International Emissions Trading Association (IETA), EcoSecurities, Climate Care, the residents of Fordside, several financial accountants at large European companies, Hockerton Housing Project and the team at BedZed.

Finally, a huge thank you is owed to my family – Francisco, Nicolas, Isabel and Emily – particularly for their patience when I arrive home from work tired and distracted. Having three young children and writing a scholarly book might seem, at first glance, somewhat incompatible, but in truth my family have both helped me focus during my working hours and provided a welcome distraction when at home.

Some parts of chapters of this book have been published previously elsewhere, and I would like to thank the following for permission to use copyrighted material: Taylor and Francis Ltd for parts of Chapter 2 from: Lovell, H. and D. M. Liverman (2011). "Understanding carbon offset technologies". *New Political Economy* **15**: 255–273 (thanks also to Professor Diana Liverman as a co-author for kindly agreeing to me reusing aspects of this paper); John Wiley & Sons Ltd for parts of Chapter 4 from: Lovell, H. (2004). "Framing sustainable housing as a solution to climate change". *Journal of Environmental Policy and Planning* **6**(1): 35–56; Emerald Group Publishing Ltd for Chapter 5 from: Ascui, F. and H. Lovell (2011). "As frames collide: making sense of carbon accounting". *Accounting, Auditing and Accountability Journal* **24**(8): 978–999 (thanks also to Francisco Ascui as the lead author of this paper for kindly agreeing to me reusing it); Cambridge University Press for parts of Chapter 6 from: Lovell, H. (2014). "Measuring forest carbon". *Governing the Climate: New Approaches to Rationality, Power and Politics*. J. Stripple and H. Bulkeley (eds). Cambridge, Cambridge University Press: 175–196; Taylor and Francis Ltd for parts of Chapter 7 from: Lovell, H. (2013). "Climate change, markets and standards: the case of financial accounting". *Economy and Society* **43**(2): 260–284.

ABBREVIATIONS

AASB	Australian Accounting Standards Board
ACCA	Association for Certified Chartered Accountants
AFOLU	Agriculture, Forestry and Other Land Use (Guidelines)
ANC	Autorité des Normes Comptables (French financial accounting standard setter)
BRE	Building Research Establishment
BSI	British Standards Institute
CAT	Centre for Alternative Technology
CCBA	Climate Community and Biodiversity Alliance
CCI	Clinton Climate Initiative
CCS	Carbon Capture and Storage
CDM	Clean Development Mechanism
CDP	Carbon Disclosure Project
CER	Certified Emission Reduction
CESP	Community Energy Saving Programme
CFL	Compact fluorescent light bulb
CHP	Combined Heat and Power
CIFOR	Centre for International Forestry Research
COP	Conference of the Parties
DCLG	Department for Communities and Local Government
DEFRA	Department for Environment, Food and Rural Affairs
DETR	Department of the Environment, Transport and the Regions
DTI	Department of Trade and Industry
DTLR	Department for Transport, Local, Government and the Regions
EFRAG	European Financial Advisory Group
ENDS	Environmental Data Services
ESRC	Economic and Social Research Council (UK)
EST	Energy Saving Trust
ETS	Emissions Trading Scheme(s)

EU ETS	European Union Emissions Trading Scheme
FASB	Financial Accounting Standards Board (US)
FCPF	Forest Carbon Partnership Facility
FoE	Friends of the Earth
GAAP	Generally Accepted Accounting Principles
GBC	Green Building Council
GEO	Group on Earth Observations
GFOI	Global Forest Observations Initiative
GHGs	Greenhouse gases
GOFC-GOLD	Global Observation of Forest and Land Cover Dynamics
GPG	Good Practice Guidance
GRI	Global Reporting Initiative
GWP	Global warming potential
HA	Housing Association
HECA	Home Energy Conservation Act
HFCs	Hydrofluorocarbons
IAS	International Accounting Standard
IASB	International Accounting Standards Board
IEAF	International Energy Accounting Forum
IETA	International Emissions Trading Association
IFRIC-3	International Financial Reporting Interpretations Committee Interpretation 3: Emission Rights
IFRS	International Financial Reporting Standards Foundation
IIED	International Institute for Environment and Development
IPCC	Intergovernmental Panel on Climate Change
IPCC GPG	Intergovernmental Panel on Climate Change Good Practice Guidance for land use, land-use change and forestry
LCA	Life cycle analysis
LULUCF	Land Use, Land-Use Change and Forestry
MRV	Measurement, reporting and verification
N_2O	Nitrous oxide
NGOs	Non-governmental organisations
ODPM	Office of the Deputy Prime Minister
OFGEM	Office of Gas and Electricity Markets
PFCs	Perfluorocarbons
PV	Photovoltaic
RCEP	Royal Commission on Environmental Pollution
RC-UK	Research Councils United Kingdom
REDD+	Reducing Emissions from Deforestation and Forest Degradation (+ = incorporating the role of conservation, sustainable management of forests and enhancement of forest carbon stocks)

RSL	Registered Social Landlord
SBSTA	Subsidiary Body on Scientific and Technical Advice
SDC	Sustainable Development Commission
SF_6	Sulphur hexafluoride
SHQS	Scottish Housing Quality Standards
STS	Science and Technology Studies
TCPA	Town & Country Planning Association
UN	United Nations
UN COP	United Nations Conference of the Parties
UNEP	United Nations Environment Programme
UNFCCC	United Nations Framework Convention on Climate Change
UN REDD	United Nations Reducing Emissions from Deforestation and Forest Degradation
VCS	Verified Carbon Standard
VER	Verified or Voluntary Emission Reduction
WBCSD	World Business Council for Sustainable Development
WRI	World Resources Institute
WWF	World Wide Fund for Nature

1

INTRODUCTION

It was whilst I was doing fieldwork in connection with my research on low carbon housing that I conducted a particularly memorable interview with a leading campaigner, who explained to me regarding a newly built and high profile zero carbon housing development in the UK: "It is a total unmitigated disaster. If it ever got out how much it cost to build that development it would put sustainability back twenty years". It was a powerful and revealing observation about the integral role of money and finance in developing solutions to the problem of climate change. His comments illustrate how low carbon economies are about politics and emotions, as well as profit margins and technological innovation. Making low carbon economies is a messy process and an intriguing one, ripe for investigation. It was his observation that set me on a pathway to writing this book. It is about the new practices and approaches that have developed over two decades of effort at climate change mitigation: a period of active construction of low carbon markets and economies. *The Making of Low Carbon Economies* examines society's response to climate change by exploring a diverse array of empirical sites – from remote sensing observations of the carbon in tropical forests to financial accounting and residential district heating – where climate change is being incorporated into everyday lives. The book describes a process of stitching climate concerns into the discourse and practices of already existing markets and economies, as well as creating new ones. *The Making of Low Carbon Economies* uses three interdisciplinary themes – heterogeneous networks, framing and commensuration – drawn variously from sociology, human geography, Foucauldian scholarship and Science and Technology Studies (STS), to explore questions about how, where and why low carbon economies are being made.

Aims and objectives

The Making of Low Carbon Economies seeks to combine theoretical explorations with detailed and up-to-date empirical research. The book joins a growing area of interdisciplinary social science research on climate change, including an increasing number of critiques of the dominant economic framing of climate change (Lohmann 2005; Spash 2007; Prudham 2009; Andrew *et al.* 2010). It is, however, not a straightforward critique of the prevalent market-orientated, neoliberal response to climate change. Rather, it attempts to promote a more nuanced and detailed approach to understanding the diverse ways in which climate change has been conceived of, and responded to, as an economic problem.

The purpose of *The Making of Low Carbon Economies* is to:

- provide an interdisciplinary perspective on low carbon economies and markets, using a mix of ideas drawn from the social sciences; and
- demonstrate the value of this broad perspective across a diverse range of empirical and topical case studies of forests, financial accounting, carbon offsets, district heating and housing.

The three interdisciplinary themes of the book – heterogeneous networks, framing and commensuration – are used to explore in-depth case studies of responses to climate change. I examine the diversity of practices, ideas, institutions and objects that have been enrolled into, and framed as, low carbon economies and markets. The empirical chapters (2 to 7) each examine a particular site (professional arenas, as well as locations) where carbon as a commodity, and climate change as a policy issue, has manifested. I evaluate the process of making discrete new markets centred on climate change – most notably, carbon markets – as well as the intrusion of climate change into already existing economies. I aim to show, through case studies, how low carbon economies have come into being in these different arenas, and what this means in practical, as well as conceptual, terms. I consider how broad social science ideas about economies and markets, coming from outside the discipline of mainstream (neoclassical) economics, inform our understanding of climate change, through concentrating on three core interdisciplinary themes, as follows.

The Part I theme of heterogeneous networks is about the dynamic, mixed assemblages of humans and material entities that comprise our world. Part I of the book considers how sociotechnical theories and concepts that are (broadly defined) about heterogeneous networks (i.e. networks of humans and non-humans), including *agencement*, actor-network theory and sociotechnical transitions theory, have been applied to markets and economies, in order to further our understandings of them as social, as well as material, structures (Callon 2007).

Part II of the book uses the concept of framing to explore the ways in which the problem of climate change is viewed, discussed and resolved. Consideration of framing is especially relevant to complex new problems such as climate change. Framing is a concept that has its origins within Foucauldian scholarship, as well as public policy analysis, but which has also been used and further developed within economic sociology and STS (Rein and Schon 1993; Abolafia 1998; Hoffman and Ventresca 1999; Laws and Rein 2003; Lovell 2004; Lohmann 2005; Callon 2007; Lohmann 2009). Part II of the book introduces and explores a number of insights that framing provides for the study of (low carbon) economies and markets, integrating ideas from across these different disciplines.

Part III of the book is about commensuration, defined as creating equivalence (Espeland and Stevens 1998) or simply "making things the same" (MacKenzie 2009). Commensuration is not an area of climate change research that has to date been given much attention (albeit with notable exceptions, see Levin and Espeland 2002; Lohmann 2009; MacKenzie 2009), and Part III of *The Making of Low Carbon Economies* aims to rectify this. Commensuration is important for climate change, particularly regarding carbon markets, because of how these discrete new markets have centred on the creation of carbon as a novel and fungible commodity, in the process bringing together different greenhouse gas emission reductions across different locations, as part of an ambitious, highly technical and inherently political project to create equivalence (MacKenzie 2009).

Who is this book for?

The Making of Low Carbon Economies is for social scientists who wish to broaden their empirical knowledge of climate change mitigation activities and who are interested in exploring interdisciplinary approaches to understanding climate change. It is for postgraduate and final year undergraduate students specialising in climate change, who wish to better comprehend and appreciate the multitude of ways in which low carbon economies are being made. I have attempted to write the book in such a way that it also has appeal to the interested non-specialist lay reader, who is coming new to the conceptual and empirical material explored within this book.

The Making of Low Carbon Economies comprises a set of empirical case studies, arranged so that readers can dip in and out or read through from start to finish if they wish. In other words, each chapter consists of analysis of a case study that is written to stand alone, but is also linked to the other chapters via the three themes of the book: Part I – heterogeneous networks; Part II – framing; and Part III – commensuration.

A note of caution: this book is not for those looking for an in-depth single-discipline account of climate change mitigation, be it from the field of human geography, sociology, politics or STS. *The Making of Low Carbon Economies* is deliberately interdisciplinary. There are three principal sets of theory that the book draws upon: economic sociology, STS and Foucauldian scholarship. A detailed account of these bodies of theory is not provided here (instead refer to Smelser and Swedberg (2010) for an overview of economic sociology and in particular,

MacKenzie *et al.* (2007) for the performativist school; Jasanoff *et al.* (1995) for an introduction to STS; and Rabinow (1984) for an edited summary of Foucault's writing and lectures, plus Dean (1999) for an overview of governmentality, and Li (2007) for an excellent application of Foucault's ideas), rather the theories are introduced as appropriate in the introductions to Parts I, II and III. The three themes of the book – heterogeneous networks, framing and commensuration – have been deliberately chosen because of their interdisciplinary reach: their ability to weave together similar ideas across disciplines that are especially pertinent to the study of low carbon economies and markets, and in a manner that aims to avoid unnecessary repetition.

Key terms

The term 'low carbon economy' is central to this book. It is a term that has increased in popularity of late, and, rather like the invoking of 'community' in climate change policy and elsewhere (Aiken 2012; Seyfang and Haxeltine 2012), it is hard to disagree within in principle, conveying as it does a sense of the possibility of combining economic growth with climate change mitigation (the classic ecological modernist 'win-win' scenario; see Andersen and Massa 2000; Murphy 2000). 'Low carbon economy' is also aided in this regard by its rather nebulous definition, which typically refers both to climate mitigation underway in particular 'green' sectors – renewable energy, energy efficiency and 'clean tech' – as well as more general economy-wide shifts, such that "...all products and services, right through the supply chain, embrace a low carbon approach" (Scottish Government 2010: 3). Widespread use of the term 'low carbon economy' began around the turn of the century: the 2003 UK Energy White Paper was, for example, called "Our energy future – creating a low carbon economy" (DTI 2003), and the first report of the UK's Climate Change Committee (2008) was called "Building a low-carbon economy: the UK's contribution to tackling climate change", plus the Stern Report (2006) has a Chapter (23) on "Supporting the transition to a low-carbon global economy". The 2003 UK Energy White Paper defines a low carbon economy as one

> ...where higher resource productivity – producing more with fewer natural resources and less pollution – will contribute to higher living standards and a better quality of life ... [where there is an] opportunity to develop, apply and export leading-edge technologies, creating new businesses and jobs. And the opportunity to lead the way, in Europe and internationally, in developing environmentally sustainable, reliable and competitive energy markets that will support economic growth in every part of the world.
>
> (DTI 2003: 6)

The Scottish Government has a somewhat more succinct and socially orientated definition of a low carbon economy as:

> . . . one where all products and services, right through the supply chain, embrace a low carbon approach. "Low carbon" is a way of thinking, behaving and operating that minimises carbon emissions while enabling sustainable use of natural resources, economic growth and quality of life improvements.
>
> (Scottish Government 2010: 3)

These definitions encompass ideas about economic growth and competitiveness. They are also focused on the national (UK, Scottish) economy, conceiving of a low carbon economy as territorially defined.

This book is not, however, about this popular definition and understanding of a low carbon economy, despite its current dominance within climate change policy and practice. It is instead about the multiple ways that climate change is being made economic: it is about 'low carbon economies' in the plural rather than the singular, embracing a broader understanding of what economies comprise and the possibility of the existence of diverse (low carbon) economies, each with different meanings, objectives and practices (see Gibson-Graham 2008) – a key point for the book, returned to later.

Before proceeding, two other, important, more generic terms underlying the book also need to be clarified – notably, 'market' and 'economy'. 'Market' is, in its traditional (neoclassical economics) guise, a much more precise term than 'economy', referring to a specific mechanism and/or location of exchange of a commodity; of buying and selling. Fligstein (1996: 658) thus defines markets as ". . . situations in which some good or service is sold to customers for a price that is paid in money (a generalised medium of exchange)". But from a sociological perspective, this definition problematically ignores the social and cultural aspects of markets. Fligstein later comes to define markets with social issues at their core, as "social structures characterised by extensive social relationships between firms, workers, suppliers, customers, and governments" (Fligstein and Dauter 2007: 105), and Zelizer (1988: 618) proposes a similarly broad definition of a market as "the interaction of social, cultural and economic factors . . . [wherein] the market is analysed as one category of social relations, which involves consumption, production and exchange under a variety of cultural and structural settings". Çalışkan and Callon add a sociotechnical angle to these social conceptualisations of a market, in which a market is defined as ". . . an arrangement of heterogeneous constituents that deploys . . . rules and conventions; technical devices; metrological systems; logistical infrastructures; texts, discourses and narratives . . . technical and scientific knowledge . . . as well as the competencies and skills embodied in living beings" (Çalışkan and Callon 2010: 3).

This broadening of the meaning of 'market' by social scientists leads to closer overlap with the term 'economy', such that they are hard to neatly distinguish. One widely accepted point of difference between an economy and a market is the territoriality of an economy, where in its modern usage an economy is taken to mean "*the state of a country or region* in terms of the production and consumption of goods and services and the supply of money" (Oxford Dictionary 2014 [emphasis added]). We observe this, for instance, in the definitions of a low carbon economy introduced earlier. In seeking to distinguish between a market and an economy, Lie (1997) acknowledges that it is hard to find a comprehensive definition of a market and its relationship to wider economies, noting: "... the market, it turns out, is the hollow core at the heart of economics" (1997: 342). Lie's approach is to embrace these 'blurred boundaries' between markets and economies, as he explains: "I am not vigilant in demarcating markets from economies tout court; the blurred boundaries are symptomatic of different conceptualizations of markets". (1997: 342) This is the approach adopted throughout the book, acknowledging that there is a continuum ranging from a discrete market to a broader economy. In other words, within *The Making of Low Carbon Economies*, the pervasive ambiguity surrounding the terminology and meaning of the key terms 'market' and 'economy' is accepted. However, for clarity, I use the term 'economy' within this book as having a broader meaning than 'market', comprising a wider set of activities, institutions and resources associated with maintaining livelihoods and welfare, including labour, capital and materials; issues that are typically seen as integral (in modern times) to the nation state. The definition of markets used in the book is of markets as "social structures characterised by extensive social relationships . . ". (Fligstein and Dauter 2007: 105), but, in keeping with the conceptual attention of the book to materials and technologies, markets are seen as comprising "... not only humans, but also *objects and artefacts, techniques, and ideas* [operating] as agents embedded in networks of calculative relations" (Fligstein and Dauter 2007: 107 [emphasis added]). Importantly, as discussed earlier, the meaning of 'economies' within this book, used in the plural, is seen as distinct from 'economy' in the singular. As noted, an 'economy' in its conventional, modern usage is taken to mean a national, capitalist, profit-seeking entity, which is distinctly territorial, defined as "... the state of a country or region...". (Oxford Dictionary 2014). Foucault, along with other scholars (see Mitchell 2008, 2011), draws our attention to the ways in which the meaning of 'economy' has evolved over time into this particular, narrow, contemporary meaning. Foucault argues that it was not, in fact, until relatively recently that a state-orientated notion of economy was introduced into political practice and, crucially, the economy became "a field of intervention for government" (2007: 95). He demonstrates how until the seventeenth century, 'economy' was about the sound management of the household, about caring and providing for the family unit: "... this idea of the economy ... at that time only ever referred to the management of a small ensemble comprising the family and the household" (2007: 103). 'Economies' in the plural captures some of these wider meanings; it embraces the possibility of difference and heterogeneity – the 'diverse economies' explored by Gibson-Graham and other scholars (see Crewe and Gregson 1998; Leyshon *et al.* 2003), wherein ". . . [a] diverse economy framing opens up opportunities for elaborating a radically heterogeneous economy and theorizing economic dynamics that foster and strengthen different economies" (Gibson-Graham 2008: 618).

The structure of the book

This book explores and furthers our understanding of the making of low carbon economies through several in-depth case studies, as follows.

Part I: Heterogeneous networks

- carbon offsets (Chapter 2)
- district heating (Chapter 3)

Part Two: Framing

- low carbon housing (Chapter 4)
- carbon accounting (Chapter 5)

Part Three: Commensuration

- forest carbon markets (Chapter 6)
- financial accounting (Chapter 7)

The chapters are briefly summarised, as follows.

Part I: Heterogeneous networks

Chapter 2: Making carbon offsets

In this chapter, I attempt to unpack the 'black box' of carbon offsetting through a critical examination of the technologies and techniques that create carbon credits. Drawing on empirical research of compliance (Clean Development Mechanism) and voluntary carbon offset markets, the diversity of technologies, techniques and devices involved in carbon offsetting is explored, ranging from refrigerant plants to systems of calculation and audit. These purpose-built carbon markets are heterogeneous networks, comprising humans and institutions, as well as technologies and things. Using examples, the considerable effort involved in making and sustaining carbon markets is outlined. The role of existing knowledge and expertise in forging and framing the two main types of carbon offset market – compliance and voluntary – is explored: discourse and practices from finance, accounting and international development have all been used to create and maintain the new carbon commodity.

Chapter 3: Implementing residential district heating

In this chapter, the implementation of district heating in two residential urban areas in Scotland is used to demonstrate how low carbon economies are made and remade. The complex network of technologies (radiators, pipes, insulation, boilers, meters), institutions, financial models and householder energy practices integral to the making of a new low carbon district heating economy is identified, and the relations between network components explored. The implementation of new district heating systems has been enabled by a UK Government climate change policy, enacted by energy utilities: the Community Energy Saving Programme (CESP). But the climate change rationale of CESP is interwoven with the pre-existing UK policy landscape of fuel poverty alleviation, welfare reform and modernisation of social housing. A heterogeneous network approach illuminates the fragility of the relations between the people and artefacts that comprise markets, showing how they can easily unravel when key actors are no longer in place or not working as originally envisaged.

Part II: Framing

Chapter 4: Low carbon housing in the UK

Chapter 4 explores a particular period in UK housing, energy and climate change policy – the turn of the century – when there was a reframing of existing sustainable housing as 'low carbon'. Until the late 1990s, the UK Government and other mainstream institutions had little interest in a longstanding community active in building sustainable homes since the 1970s. What changed was the rise of the problem of climate change on the UK's policy agenda. Existing sustainable housing provided a potential 'ready-made' low carbon solution. Chapter 4 is about the politics of framing and reframing. It also demonstrates how materials and technologies are a key component of frames and explores some of the difficulties of using low carbon materials and technologies discursively – i.e. as a way to bolster arguments in favour of climate change mitigation.

Chapter 5: The frames of carbon accounting

Chapter 5 also draws on theories of framing to help explain the divergent understandings and practices currently encompassed by the term 'carbon accounting'. The empirical core of the chapter is based on a review of contemporary problems in carbon accounting, such as inconsistencies in the metrics used to calculate the global warming potential of greenhouse gases and failure to agree on international standards for carbon financial accounting. Tensions and contradictions in carbon accounting can be understood as the result of 'collisions' between at least five overlapping discursive frames – namely, physical, political, market-enabling, financial and social/environmental modes of carbon accounting. These multiple frames of carbon accounting reflect the relatively recent formation of a new field of expertise and are an expected outcome of the complex process of forging new kinds of low carbon economy.

Part III: Commensuration

Chapter 6: Measuring and standardising forest carbon for REDD+

Chapter 6 investigates the extension of carbon market rules, norms and practices of measurement to tropical forests. Such acts of commensuration are shown to be mostly invisible to the public, but nonetheless have a significant influence on the overall effectiveness of forest carbon markets and other low carbon economies. The policy ambition of accurately measuring forest carbon (known as 'MRV', for 'measurement, reporting and verification') has become a dominant feature of the making of a market for the carbon stored in forests. The science of remote sensing is at the forefront of attempts to make forests manageable through MRV and therefore stable within international carbon markets. This chapter explores the development of new techniques designed to make forests work within carbon markets, including remote sensing expertise and technologies, as well as standards.

Chapter 7: The treatment of carbon in financial accounts

Chapter 7 investigates the social, political and organisational aspects of the treatment of carbon credits in financial accounts. Since the withdrawal of international accounting guidance (IFRIC-3) in 2005, there has been no recommended financial accounting standard for carbon credits, and a diversity of practices has emerged. Carbon sits between and challenges a number of International Accounting Standards (IAS), including 20, 38 and 39. Chapter 7 provides a detailed empirical case study into how the large-scale experiment of carbon markets has manifested within a particular area of professional expertise – financial accounting. A focus on standards usefully draws attention to the hidden work these and other acts of commensuration do and helps to explain why some things such as emission allowances are difficult to standardise, providing insights into the 'coping strategies' employed when there is a prolonged absence of standards.

Chapter 8: Summary and conclusions

This final chapter reflects on a core ambition of the book: to investigate the diversity of ways in which climate change is being understood within markets and economies. It ends the book on a relatively optimistic note, discussing how the empirical chapters of the book show that a significant shift in our capabilities and response to climate change is underway. This shift is evident when we examine the development of low carbon economies and markets across many different sites. If we 'lift the stone' on such topics as house building in the UK, provision of energy in cities and financial accounting, we observe how carbon as a commodity, and climate change as an issue, are gradually being woven, often in quite subtle ways, into the daily lives and decision-making of actors within heterogeneous markets and economies.

Bibliography

Abolafia, M. Y. (1998). "Markets as cultures: an ethnographic approach". *The Laws of the Markets*. M. Callon (ed.). Oxford, Blackwell Publishers/The Sociological Review: 69–85.

Aiken, G. (2012). "Community transitions to low carbon futures in the Transition Towns Network (TTN)". *Geography Compass* **6**(2): 89–99.

Andersen, M. S. and I. Massa (2000). "Ecological modernization – origins, dilemmas and future directions". *Journal of Environmental Policy and Planning* **2**(4): 337–345.

Andrew, J., M. A. Kaidonis and B. Andrew (2010). "Carbon tax: challenging neoliberal solutions to climate change". *Critical Perspectives on Accounting* **21**(7): 611–618.

Çalışkan, K. and M. Callon (2010). "Economization, Part II: a research programme for the study of markets". *Economy and Society* **39**(1): 1–32.

Callon, M. (2007). "What does it mean to say that economics is performative?" *Do Economists Make Markets? On the Performativity of Economics*. D. MacKenzie, F. Muniesa and L. Siu (eds). Princeton, NJ, Princeton University Press: 311–357.

Crewe, L. and N. Gregson (1998). "Tales of the unexpected: exploring car boot sales as marginal spaces of contemporary consumption". *Transactions of the Institute of British Geographers* **23**(1): 39–53.

Dean, M. (1999). *Governmentality: Power and Rule in Modern Society*. London, Sage Publications Ltd.

Department of Trade and Industry (DTI) (2003). "Our energy future – creating a low carbon economy: Energy White Paper". Online at: http://webarchive.nationalarchives.gov.uk/20090609003228/www.berr.gov.uk/files/file10719.pdf (accessed 14 August 2014).

Espeland, W. N. and M. L. Stevens (1998). "Commensuration as a social process". *Annual Review of Sociology* **24**: 313–143.

Fligstein, N. (1996). "Markets as politics: a political-cultural approach to market institutions". *American Sociological Review* **61**(4): 656–673.

Fligstein, N. and L. Dauter (2007). "The sociology of markets". *Annual Review of Sociology* **33**: 105–128.

Foucault, M. (2007). *Security, Territory, Population*. Basingstoke, Palgrave Macmillan.

Gibson-Graham, J.-K. (2008). "Diverse economies: performative practices for other worlds". *Progress in Human Geography* **32**(5): 613–632.

Hoffman, A. J. and M. J. Ventresca (1999). "The institutional framing of policy debates: economics versus the environment". *American Behavioural Scientist* **42**(8): 1368–1392.

Jasanoff, S., G. E. Markle, J. C. Petersen and T. J. Pinch (eds) (1995). *Handbook of Science and Technology Studies*. London, Sage.

Laws, D. and M. Rein (2003). "Reframing practice". *Deliberative Policy Analysis: Understanding Governance in the Network Society*. M. A. Hajer and H. Wagenaar (eds). Cambridge, Cambridge University Press: 172–206.

Levin, P. and W. N. Espeland (2002). "Pollution Futures: Commensuration, Commodification, and the Market for Air". *Organizations, Policy, and the Natural Environment: Institutional and Strategic Perspectives*. A. J. Hoffman and M. J. Ventresca (eds). Stanford, CA, Stanford University Press: 119–147.

Leyshon, A., R. Lee and C. C. Williams (eds) (2003). *Alternative Economic Spaces*. London, Sage.

Li, T. M. (2007). *The Will to Improve: Governmentality, Development, and the Practice of Politics*. Durham, NC, and London, Duke University Press.

Lie, J. (1997). "Sociology of markets". *Annual Review of Sociology*: 341–360.

Lohmann, L. (2005). "Marketing and making carbon dumps: commodification, calculation and counterfactuals in climate change mitigation". *Science as Culture* **14**(3): 203–235.

Lohmann, L. (2009). "Toward a different debate in environmental accounting: the cases of carbon and cost-benefit". *Accounting, Organizations and Society* **34**(3): 499–534.

Lovell, H. (2004). "Framing sustainable housing as a solution to climate change". *Journal of Environmental Policy and Planning* **6**(1): 35–56.

MacKenzie, D. (2009). "Making things the same: gases, emission rights and the politics of carbon markets". *Accounting, Organizations and Society* **34**(3–4): 440–455.

MacKenzie, D., F. Muniesa and L. Siu (eds) (2007). *Do Economists Make Markets? On the Performativity of Economics*. Princeton, NJ, Princeton University Press.

Mitchell, T. (2008). "Rethinking economy". *Geoforum* **39**: 1116–1121.

Mitchell, T. (2011). *Carbon Democracy: Political Power in the Age of Oil*. New York, NY, Verso Books.

Murphy, J. (2000). "Editorial – ecological modernisation". *Geoforum* **31**(1): 1–8.

Oxford Dictionary (2014). "Definition of 'economy'". Online at: www.oxforddictionaries.com/definition/english/economy (accessed 4 June 2014).

Prudham, S. (2009). "Pimping climate change: Richard Branson, global warming, and the performance of green capitalism". *Environment and Planning A* **41**(7): 1594–1613.

Rabinow, P. (ed.) (1984). *The Foucault Reader: An Introduction to Foucault's Thought*. St Ives, Clays Ltd.

Rein, M. and D. Schon (1993). "Reframing policy discourse". *The Argumentative Turn in Policy Analysis and Planning*. F. Fischer and J. Forester (eds). London, UCL Press Ltd.: 145–166.

Seyfang, G. and A. Haxeltine (2012). "Growing grassroots innovations: exploring the role of community-based initiatives in governing sustainable energy transitions". *Environment and Planning C: Government and Policy* **30**: 381–400.
Smelser, N. J. and R. Swedberg (2010). *The Handbook of Economic Sociology*. Princeton, NJ, Princeton University Press.
Spash, C. L. (2007). "The economics of climate change impacts à la Stern: novel and nuanced or rhetorically restricted?" *Ecological Economics* **63**(4): 706–713.
Stern, S. N. (2006). "The Stern Review on the economics of climate change". London, HM Treasury.
Scottish Government (2010). "Towards a low carbon economy for Scotland: discussion paper". Glasgow, Scottish Government.
Zelizer, V. (1988). "Beyond the polemics on the market: establishing a theoretical and empirical agenda". *Sociological Forum* **3**(4): 614–634.

Part I

HETEROGENEOUS NETWORKS

Heterogeneous networks is an umbrella term that usefully spans a number of related and overlapping concepts and theories with a shared interest in sociotechnical relations – i.e. networks of human and non-human entities – including materiality, actor-network theory, sociotechnical transitions theory, *agencement*, market devices, relational economies and governmentality. In this introduction to Part I, I explain the theme of heterogeneous networks and explore its origins. I also look ahead to the chapters within Part I which use the idea of heterogeneous networks to explore how new markets have been made centred on the commodity of greenhouse gas emission reductions (Chapter 2 on carbon markets), and how existing energy markets and economies have been recast through the introduction of a new heating technology: district heating (Chapter 3).

The main theoretical insights provided by the theme of heterogeneous networks, as applied to markets and economies, are as follows:

- Markets and economies are made up of a mix of actors – human and non-human.
- The empirical focus of heterogeneous network scholarship is this mix of actors, as well as the relations that hold them together.
- A heterogeneous network conceptualisation deliberately lays bare the 'nuts and bolts' of markets and economies – their component parts – and aims to demystify them.
- Markets and economies appear to be stable and 'locked-in', but there is constant work 'behind the scenes' in maintaining relations.
- Heterogeneous networks can therefore easily disintegrate if one or two key objects or people are removed, leave or act in unforeseen ways.
- A heterogeneous network approach gives confidence to decision-makers and those in positions of authority that markets and economies can be designed and made from scratch.
- New markets are experimental and, in order to encourage learning, should continue to be seen as experimental long after their initial set-up.

- Market design requires a vision for the whole system, as well as the mechanisms (e.g. calculative devices) to do the practical work of bringing it into being.
- If a market is not well-designed, then 'overflowing' occurs (i.e. the diverse and inherently fragile network of things, ideas, institutions and practices that comprise a heterogeneous network start to unravel and fail to function as originally planned, within its pre-designated market boundaries).

A heterogeneous network is a dynamic and fragile sociotechnical network comprising a mix of humans and non-humans (objects, technologies and so on). As Law (1992: 2) explains, the origins of the concept come from a recognition that "... the social is nothing other than patterned networks of heterogeneous materials" and that "... these networks are composed not only of people, but also of machines, animals, texts, money, architectures ...". Whilst this definition is not focused on markets and economies per se, there are scholars who have applied the idea of a heterogeneous network to this realm (albeit using slightly different terminology), such as Barry and Slater (2002: 180), who observe that: "Economic processes can themselves be treated as just another kind of socio-technical-discursive arrangement, like cars or nuclear physics ... comprising heterogeneous actors whose properties emerge from specific but contested material arrangements"; and Munro and Smith, who describe markets as "... assemblages of people, tools, equipment, technical devices, algorithms and so on which are actively arranged but which partly arrange themselves" (2008: 359). The 'assemblages' described by Munro and Smith are a translation of the French term *agencement*, introduced into economic sociology by the performativist school of economic sociology (see MacKenzie *et al.* 2007). As explained later in this chapter, the term *agencement* is borrowed from Deleuze and Guattari (2004)), specifically "... to convey the idea of a combination of heterogeneous elements that have been carefully adjusted to one another ..." within markets and economies (Callon 2007b: 319). A heterogeneous network approach pays close attention to the processes of enrolling, stabilising and "rendering passive" the multiple (and inherently fragile) entities that comprise markets and economies (Çalışkan and Callon 2010: 6; Hardie and Mackenzie 2007).

The term heterogeneous network has been chosen for Part I of *The Making of Low Carbon Economies* because it draws together sociological conceptualisations of markets and economies with STS ideas about materiality (i.e. the influential role of things and technologies in structuring social life). Here, I consider five key insights provided by a heterogeneous network approach to the study of markets and economies that are especially pertinent to the case studies examined in Chapters 2 and 3.

First, a heterogeneous network approach is useful in specifying that markets and economies can and should be empirically investigated by seeking to identify their component parts: the technologies, ideas, bodies of expertise, individuals and institutions that they comprise. The concept of heterogeneous networks is thus important because it extends the subject of research beyond individuals, organisations and institutions to encompass the material substance of markets and economies: their technologies, gadgets, material sites of production and consumption and so on. This conceptualisation of markets and economies stems from the idea of materiality – a foundational concept within STS, which is about society involving not only human actors and social relations, but also artefacts and other non-human things (e.g. trees – see Chapter 6) (Bijker *et al.* 1987; Jasanoff *et al.* 1995; Pinch and Swedberg 2008). STS scholars take materiality as their starting point for social science research, with a belief that:

> . . . in order to understand domination [power relations] we have to turn away from an exclusive concern with social relations and weave them into a fabric that includes non-human actants, actants that offer the possibility of holding society together as a durable whole.
>
> (Latour 1991: 103)

So, attention to the materiality of social relations is really at the heart of the concept of heterogeneous networks, with its focus on *socio*technical relations and the belief (albeit controversial to some; see, for example, Collins and Yearley 1992) that non-human actors can have agency attributed to them. An example of this close attention to the role of materials within markets and economies and their capacity for agency is the notion of 'market devices', referring to

> the material and discursive assemblages that intervene in the construction of markets . . . From analytical techniques to pricing models, from purchase settings to merchandising tools, from trading protocols to aggregate indicators, the topic of market devices includes a wide array of objects.
>
> (Muniesa *et al.* 2007: 2)

These devices play an important role within markets and economies, because of their ability to ". . . render things, behaviours and processes economic" (2007: 3) and facilitate market stability through ". . . configuring economic calculative capacities and in qualifying market objects" (2007: 5).

For low carbon economies, this attention to things and materials is particularly important, simply because many of the technologies and materials that make up low carbon economies are new. It is therefore more likely that they will fail or will not function as planned, thereby fundamentally destabilising or altering markets and economies. We see this, for example, in Chapter 3, where new household metering technologies failed to provide accurate data on individual household heat consumption in a way anticipated by the City Council, thereby putting pressure on the economic viability of district heating. A heterogeneous networks approach also forces us to recognise how even totally new markets designed to mitigate climate change – the international carbon markets explored in Chapter 2 – have, in effect, been carved out of existing technologies, things, expertise and ways of doing, which continue to have influence. Indeed, the existing 'stuff' – pipes, wires, buildings, institutions – that was in place before climate change became a problem can result in a situation of 'lock-in' to this existing unsustainable infrastructure (Unruh 2002) and is one reason why progress to mitigate climate change has been slow.

There are valuable insights provided here, too, by Foucault's notion of governmentality through its connection between ways of thinking and bodies of knowledge ('rationalities') and how they are operationalised and governed through 'technologies' (defined as instruments, practices and techniques)[1] (Dean 1999). Governmentality is thus an approach that is inherently sociotechnical. A heterogeneous networks lens draws our attention to the practical limitations of the rationalities of governmentality, depending on the technologies available. Conversely, the emergence of novel technologies allows opportunities for new rationalities to come into being. For example, Foucault sees what he terms "disciplinary technologies" (such as statistics, data, auditing; see Rabinow 1984: 17) as a precursor to capitalism, because these technologies allowed the economy to be framed in a different way, as for the first time new, reliable national information and data was produced that went significantly beyond the household to encompass whole populations, resource statistics, monetary flows and so on (Lemke 2002; Mitchell 2011). In other words, the emergence of these new types of disciplinary technologies – statistical data about whole populations – was vital to the rise of capitalism, as Rabinow (1984: 18) explains:

> The growth and spread of disciplinary mechanisms of knowledge and power preceded the growth of capitalism in both the logical and temporal sense. Although these technologies did not cause the rise of capitalism, they were the prerequisites for its success.

Mitchell echoes Foucault in his detailed historical analysis of the oil industry, describing how, according to his analysis, it was not until the 1930s that "... economists developed practical tools for measuring and managing the value of money that became part of the novel day-to-day machinery of monetary circulation that was soon to be recognised as 'the economy'" (Mitchell 2011: 135).

So whilst it might sound self-evident that a multitude of people, objects and technologies determine what markets and economies are and how they evolve, such a framing serves as a useful counter to arguments (used by practitioners within carbon markets and beyond) that it is inherently problematic to intervene in markets and economies or, in Callon's (2009: 539) terms, that markets are somewhat mysterious and incomprehensible "quasi-natural realities". A heterogeneous network approach highlights the diverse and often quite mundane things, ideas, practices and so on that comprise economies and markets, directing our critical attention toward these diverse components, as well as the relations that hold them together.

A second valuable feature of heterogeneous network conceptualisations is that they draw our attention to the relations between things and people: the 'glue' that holds markets and economies together. Although there is a semblance of stability and lock-in within heterogeneous networks, these relations are inherently fragile: there is constant work involved to keep the illusion of stability going (Graham and Marvin 2001). When key elements of markets break down, disappear or change roles, then previously hidden relations that were allowing the network to function and appear stable are laid bare. These moments of change are of interest to us as researchers – an empirical research opportunity – because of what they reveal. Such ideas about change and the inherent fragility of sociotechnical relations are captured within the STS concept of actor-network theory, as well as its economic sociology variant – *agencement* or assemblage – two concepts that will now be briefly introduced in turn.

Actor-networks are defined as dynamic, heterogeneous assemblages of humans and material entities (Callon 1986; Law 1992; Law and Hassard 1999). Within actor-networks, agency and power arise from the relationships between network members: agency is an outcome of networks, and thus it is only through the formation and operation of networks that change takes place. Further, by assuming no *a priori* distinction between human and non-human entities, the interactions between technologies, individuals and institutions are able to be explored in innovative ways. As Murdoch (1997: 753) states, one of actor-network theory's main advantages is that it is a challenging, creative analytical framework, thus forcing people to: "... look afresh at the categories, divisions and boundaries that frequently divert our attention away from the nonhuman multitudes which make up our world". A key feature of the relationships that comprise actor-networks is that they are largely invisible or, rather, they are such an integral feature of modern life that they have attracted little policy or theoretical attention, a phenomenon termed 'black boxing' (Hinchliffe 1996; Graham and Marvin 2001), whereby, for example, infrastructures providing energy and transport "... gradually diffuse to become taken for granted and 'normalised' as essential, but largely invisible, supports of modern urban life" (Graham and Marvin 2001: 180).

In actor-networks change occurs through the construction or dissolution of relationships within the network. Moreover, the identities and agency of actors are defined through their network relations. As noted, power arises from these relations and not the entities themselves, and change is therefore catalysed only by collective action (Holloway 1998; Murdoch 2001). As with sociotechnical systems (see later in this chapter), there are various stages in the formation of an actor-network. A coherent actor-network emerges when the process of 'translation' – whereby diverse entities are enrolled into a network (Callon 1986) – is complete. In other words, when the heterogeneous elements of the network – ranging from humans to technologies and animals – are redefined so they behave in accordance with others in the network thus: "... despite the heterogeneous quality of any previous identities these entities now work in unison, thereby enabling the enrolling actor (the 'centre') to 'speak' for all" (Murdoch 1998: 362).

Because of the inherent fragility and dynamism of actor-networks, material objects play an important role in conferring stability (Latour 1991; Law 2001). Moments of change or breakdown are of particular relevance with regard to climate change, because as a new, big problem it has the potential to disrupt, and is already disrupting, existing ways of doing: it is a crisis that has the capacity to break down and expose the underlying relations of markets and economies. This is evident in Chapter 3, for example, where the installation of a climate mitigation measure in a housing development – new district heating – ran into problems when key people and technologies disappeared or failed to function as planned. The case study in Chapter 3 tells us something about the particularities of implementing district heating, but also provides insights into the processes by which wider housing and energy economies operate and respond to change.

Attention to the relations between things and people – the 'glue' that holds markets and economies together – is also the object of attention of a concept closely related to actor-network theory – that of *agencement* (Callon 2007a, 2007b; Hardie and Mackenzie 2007). Callon borrowed the term *agencement* from Deleuze and Guattari (2004) and defines *agencements* as "... arrangements endowed with the capacity of acting in different ways depending on their configuration" (Callon 2007b: 320). Hardie and MacKenzie (2007: 59) further clarify that an *agencement* is "... the arrangement (in the broadest sense) of people, technical systems and so on that constitutes [a market]". *Agencements* are not just a static arrangement of people, things, materials and their meanings, but they have, collectively, and as a matter of necessity and routine, the capacity to act. In other words, they are "... assemblages ... which are actively arranged but which partly arrange themselves" (Munro and Smith 2008: 359). The concept of *agencement* is hence closely aligned with actor-network theory, through a shared conceptualisation of heterogeneous network-based forms of agency.

Agencement has been developed as a term mostly within a branch of economic sociology specifically about markets and economies, called the performativist school (see MacKenzie 2003; Fligstein and Dauter 2007; MacKenzie *et al.* 2007). Performativity is partly about the role of (economics) expertise in shaping how particular markets and economies function (MacKenzie 2006b; see later in this chapter), but it is also about the role of materials and technologies within *agencements*, and how they do not always work as planned, as Callon (2007: 17) explains:

> . . . the notion of . . . performativity goes beyond human minds and deploys all the materialities comprising the socio-technical *agencements* that constitute the world in which these agents are plunged: performativity leaves open the possibility of events that might refute, or even happen independently of, what humans believe or think.

The relations between materials and people are also central to an expanding area of inquiry in economic geography, termed 'relational economic geography', which integrates economic, political, social and cultural issues within the study of economies and markets, positing that economies can only be explained with reference to feelings, emotions and other 'non-economic' goals (see Jones 2013 for a review). Such an approach is in keeping with the theme of heterogeneous networks and marks a strong contrast to more 'traditional' economic geography, with its core focus on regions and other spatial representations (Bathelt 2006). Relational economic geography embraces a small-scale, bottom-up approach to understanding processes of economic development, one which "emphasizes its contextual, path-dependent and contingent nature" (Jones 2013: 3) and draws on many of the ideas and concepts outlined here, such as actor-network theory and materiality. It thus provides a productive avenue for considering the diverse low carbon economies that are examined in the empirical chapters of this book (see also Leyshon *et al.* 2003; Gibson-Graham 2006, 2008).

Third, a heterogeneous network approach makes it evident, through its interest in breaking down and demystifying network relations, that markets and economies can be made from scratch, and often are. In studying carbon markets in particular (see Chapter 2), this is a helpful reminder of the capability of organisations (public and private) to plan, create, organise and sustain new modes of low carbon production and consumption through markets and economies. We can usefully draw here on the performativist school of economic sociology (Fligstein and Dauter 2007; MacKenzie *et al.* 2007), which is interested in how markets and economies are crafted from the disciplinary script of economics. In other words, a concern with ". . . the role of economists (and others) in the creation of cultural tools *that actually enact the market* in fields like finance" (Fligstein and Dauter 2007: 118 [emphasis added]). The performative turn that underpins this particular conceptualisation of markets and economies recognises that if the economy has to be made, then it could potentially be made differently (Callon 2006; MacKenzie 2006a; Smith *et al.* 2006). This is a crucial insight for *The Making of Low Carbon Economies*, which is centrally about the possibilities of remaking markets and economies in a different way – i.e. in a way attentive to climate change. Markets and economies, characterised as *agencements*, are dynamic and inherently fragile, and there is hence the possibility of reframing and remaking how they operate (Callon 2009).

Chapter 2 examines one such market that has been developed internationally for carbon offsetting: a new market made from scratch as a direct, large-scale mitigation response to climate change. The Clean Development Mechanism (CDM), the main international carbon market established under the 1997 United Nations Kyoto Protocol, provides an excellent example of a public sector-constructed, heterogeneous network, because of the multitude of technologies, things, institutions and so on that comprise the market. The CDM provides a mechanism for the exchange of emission reductions: countries in the developing world are able to trade emission reductions resulting from projects they have purposefully set up with countries in the developed world who are looking to meet carbon emission reduction targets. Within the CDM, many things and people are required to make a carbon offset; and, as recent difficulties in the CDM market show, it is a finely crafted, delicate network of materials, institutions and ideas, vulnerable to unravelling. The contemporary debate about the long-term viability of carbon markets largely neglects the experimental element of these new markets (even despite the first phase of the European Emissions Trading Scheme being explicitly set up as experimental; see MacKenzie 2007; Callon 2009). Callon explains the value of conceiving of markets as experimental when he suggests: "[w]hat sociology and anthropology could bring to the [markets] debate is precisely a recognition of the experimental character of markets and market organisation and the need to debate the consequences of experimentation. It is a collective learning process" (Callon, quoted in Barry and Slater 2005: 114). The value of conceptualising heterogeneous carbon markets as experimental is that it offers positive possibilities for change: the rules, procedures and practices of the market are still open to debate, questioning and alteration (MacKenzie 2007). Change might come in the form of new institutions and international rules and regulations, but also via 'bottom-up' changes in practices and ways of doing – e.g. new types of carbon accounting, carbon market politics, the reclassification of emission allowances as an entity in financial accounting and so on.

A core challenge for market actors operating within these fragile and dynamic heterogeneous networks is to establish a situation of relative stability or 'lock-in': a progressive narrowing of options whereby markets become stabilised and thus able to operate. As Callon (1998: 50) explains: "Once organised and hence locked-in, the market becomes calculable by the agents". 'Lock-in' is thus seen in a positive light within the performativist school, in contrast to its more negative thrust in the consideration of sociotechnical transitions theory (see later in this chapter). For Callon, 'lock-in' is the endpoint of a process of 'becoming economic', rather than a position at the midst of struggle and change (Callon 1998). Closely associated with lock-in is the concept of framing (see Part II). Frames allow agents to calculate and make sense of the market, by drawing boundaries around what is in and out of it. Frames are not static, but are constantly evolving as parts of the market 'overflow' (Callon 1998).

A fourth insight of the heterogeneous networks concept is that design of markets and economies requires a vision for the whole system. As Mitchell (2008) argues, there is no 'pure' economy: all elements of society and the material world are mixed together within markets. Using the example of Edison's electrical system (drawing on Thomas Hughes' *Networks of Power* (1983)), Mitchell makes the case that what was required for Edison's electricity innovations to become widely used were new technical processes and forms of calculation that brought together and co-ordinated the myriad aspects of electricity systems: "electrical, chemical, economic and social" (2008: 1117) (see also the introduction to Part III of this book for discussion of commensuration – creating equivalence – within markets and economies). Callon and Muniesa's (2005) work on the constitution of markets as collective calculating devices similarly makes the point that calculation is distributed widely across the people and things of the market: it is not effected through a single price mechanism or even through some form of human agency alone (Velthius 2005; Buenza *et al.* 2006; Munro and Smith 2008). The characterisation of markets as collective calculating devices highlights how the practice of making things calculable (like that of calculation itself) is an uneven, unequal and contestable process. Further, it argues that in order for calculation to be successful it must encompass the whole market or economy; a vision is required. A new (and big) issue like climate change sits outside of already existing calculative devices and therefore is at a structural disadvantage: climate change either has to fit existing forms of market calculation or, more fundamentally, requires an upheaval in the metrics of calculation. This conceptual insight is reflected upon on in Chapter 2, where it is evident that whilst there was originally a clear vision for a discrete new carbon market (the CDM), it nevertheless has run into problems at the point of implementation, where arguably the practical work of bringing it into being, including its modes of governance and "disciplinary technologies" (Foucault 2007), had not been adequately considered at the outset. Chapter 2 provides empirical evidence to substantiate a concern of Callon's (2009) regarding carbon markets – namely, that their framing is premature. Market actors (including policymakers and politicians) have been trying to establish boundaries and rules before really understanding the issues at stake or truly appreciating the novelty of carbon as a commodity.

Heterogeneous network conceptualisations of markets and economies – notably, that of *agencement* – have tended to concentrate on the 'here and now' of market operation: the contemporary constitution of, and relations between, market actors. It is a body of theory that is hence weaker on the visions and 'big picture', longer-term drivers of change: how markets and economies significantly alter over time and why. *Agencement* is helpful in demonstrating the multitude of people and things that constitute discrete markets, but it has, to date, been applied mostly to small-scale, relatively static case studies. In other words, the majority of empirical cases examined by scholars and characterised as *agencements* are small-scale cases of particular markets (Callon 1998; Callon *et al.* 2007; Breslau 2013), which provide relatively limited insight for larger-scale reform of markets and economies, prompted, for example, by significant, global problems such as climate change. An area of scholarship that does, however, focus precisely on such issues is that of sociotechnical system transitions (Geels 2004, 2011; Smith *et al.* 2005; Markard *et al.* 2012). Sociotechnical transitions theory is designed to explain and better understand major shifts in technologies, especially radical shifts, which mark a break from past technologies and practices. This popular and rapidly expanding body of research fits within the general theme of heterogeneous networks, because of its core interest in sociotechnical relations, with a sociotechnical system defined as "...a range of elements ... linked together to achieve functionality, for example, technology, regulation, user practices and markets, cultural meaning, infrastructure, maintenance networks and production systems" (Geels 2004: 1). A sociotechnical system is typically a large infrastructure system – e.g. electricity, telecommunications or transport (Hughes 1983; Schot *et al.* 1994; Coutard 1999; Graham and Marvin 2001; Weber 2003; Geels 2004).

Although originally developed to explain examples of historical change in infrastructures, in recent years transitions theory has been increasingly applied to consider contemporary changes in large modern infrastructures, and in particular how to deliberately effect transitions in response to a growing array of environmental problems, not least climate change. Transitions are defined as "complex and long-term processes [taking place over '40–50 years'] comprising multiple actors" (Geels 2011: 1) and, in line with other STS approaches and the performativist school of economic sociology (see earlier), transitions are described as "co-evolutionary" (Schot 1992; Geels 2004) – i.e. necessarily sociotechnical: neither just about technology, nor simply about society. A subset of sociotechnical systems scholarship that is most developed is a concept called the multi-level perspective on sustainability transitions (Geels 2004, 2011; Smith *et al.* 2005, 2010), and it is the multi-level perspective that is drawn on in this book, both in Part I and beyond (and hereafter is simply referred to as 'transitions theory').

The 'multi-level' aspect of transitions theory refers to the different components where sociotechnical stability and innovation can be identified: the core middle-level regime "the locus of established practices and associated rules that stabilize existing systems" (2011: 26); the lower-level, small-scale innovation niches, where radical innovations occur, protected from regime 'lock-in'; and the most stable, high-level sociotechnical landscape, the "technical and material backdrop", where change is slowest and most difficult to effect (2011: 28). The temporal thrust of transitions theory (with its theoretical basis in evolutionary economics; see Hughes 1983) provides a useful means of exploring the ways in which climate change might disrupt existing heterogeneous networks. For what we need to understand, and learn from, is how heterogeneous networks have changed in the past in response to other big problems – variously termed 'exogenous shocks', 'critical problems' and 'reverse salients' by transitions theory scholars (Hughes 1983; Kemp 1994; Unruh 2002). So what this branch of scholarship on sociotechnical transitions usefully adds to a heterogeneous networks conceptualisation is consideration of the triggers for innovation (what provokes a big change; the catalyst), as well as a temporal or historical perspective. Issues concerning the long-term pattern of change in markets and economies surface throughout Chapters 2 and 3 – for instance, the recognition that even new low carbon economies such as international carbon markets (Chapter 2) have, in effect, been carved out of existing markets, expertise and ways of doing. Further, Chapter 2 provides an example of how markets evolve through interacting with each other, drawing on the sociotechnical transitions concept of an innovation niche.

In summary, this introduction to Part I of the book has explained the meaning of heterogeneous networks and its application within the book, as well as introducing a number of concepts drawn mostly from STS and economic sociology and geography concerned with types of heterogeneous network, including actor-network theory, materiality, governmentality, relational economies, transitions theory and *agencement*.

Note

1 The Foucauldian governmentality definition of technology denotes practices and 'techniques and devices' (e.g. audits, systems of measurement), rather than engineered material technologies (i.e. the more conventional definition – wind turbines, cooking stoves and the like). In other words, technologies are defined by governmentality scholars in an unusual way, referring mostly to 'softer' policy techniques and partnerships, whereas STS scholars attend, as one might expect, to more conventional technologies – power stations, pipes and wires; see Hughes (1983); Moss (2001). Examples of governmental technologies range from policy instruments to new tools of management and institutional devices; see Dean (1999); Oels (2005).

Bibliography

Barry, A. and D. Slater (2002). "Introduction: the technological economy". *Economy and Society* **31**(2): 175–193.

Bathelt, H. (2006). "Geographies of production: growth regimes in spatial perspective 3-toward a relational view of economic action and policy". *Progress in Human Geography* **30**(2): 223–236.

Bijker, W. E., T. P. Hughes and T. J. Pinch (eds) (1987). *The Social Construction of Techological Systems: New Directions in the Sociology and History of Technology*. Cambridge, MA, MIT Press.

Breslau, D. (2013). "Designing a market-like entity: economics in the politics of market formation". *Social Studies of Science* **43**(6): 829–851.

Buenza, D., I. Hardie and D. MacKenzie (2006). "A price is a social thing: towards a material sociology of arbitrage". *Organization Studies* **27**: 721–745.

Çalışkan, K. and M. Callon (2010). "Economization, Part II: a research programme for the study of markets". *Economy and Society* **39**(1): 1–32.

Callon, M. (1986). "Some elements in a sociology of translation: domestication of the scallops and fishermen of St. Brieuc Bay". *Power, Action, Belief*. J. Law (eds). London, Routledge and Kegan Paul: 196–233.

Callon, M. (1998). "Introduction: the embeddedness of economic markets in economics". *The Laws of the Markets*. M. Callon (ed.). Oxford, Blackwell Publishing/The Sociological Review: 1–57.

Callon, M. (ed.) (1998). *The Laws of the Markets*. Oxford, Blackwell Publishing/The Sociological Review.

Callon, M. (2006). "What does it mean to say that economics is performative? Working paper". Online at: www.csi.ensmp.fr/ (accessed 13 January 2007).

Callon, M. (2007a). "An essay on the growing contribution of economic markets to the proliferation of the social". *Theory, Culture and Society* **24**(7/8): 139–162.

Callon, M. (2007b). "What does it mean to say that economics is performative?" *Do Economists Make Markets? On the Performativity of Economics*. D. MacKenzie, F. Muniesa and L. Siu (eds). Princeton, NJ, Princeton University Press: 311–357.

Callon, M. (2009). "Civilizing markets: carbon trading between *in vitro* and *in vivo* experiments". *Accounting, Organizations and Society* **34**(3–4): 535–548.

Callon, M., Y. Millo and F. Muniesa (eds) (2007). *Market Devices*. Oxford, Blackwell Publishing.

Callon, M. and F. Muniesa (2005). "Economic markets as calculative collective devices". *Organization Studies* **26**(8): 1229–1250.

Collins, H. M. and S. Yearley (1992). "Journey into space: on epistemological evasions". *Science as Practice and Culture*. A. Pickering (ed.). Chicago, IL, University of Chicago Press: 369–389.

Coutard, O. (ed.) (1999). *The Governance of Large Technical Systems*. London, Routledge.

Dean, M. (1999). *Governmentality: Power and Rule in Modern Society*. London, Sage Publications Ltd.

Deleuze, G. and F. Guattari (2004). *A Thousand Plateaus*. Brian Massumi (trans). London, Continuum.

Fligstein, N. and L. Dauter (2007). "The sociology of markets". *Annual Review of Sociology* **33**: 105–128.

Foucault, M. (2007). *Security, Territory, Population*. Basingstoke, Palgrave Macmillian.

Geels, F. W. (2004). *Technological Transitions and System Innovations: A Co-Evolutionary and Socio-Technical Analysis*. Cheltenham, Edward Elgar.

Geels, F. W. (2011). "The multi-level perspective on sustainability transitions: responses to seven criticisms". *Environmental Innovation and Societal Transitions* **1**(1): 24–40.

Gibson-Graham, J.-K. (2006). *A Post-Capitalist Politics*. Minnesota, MN, University of Minnesota Press.

Gibson-Graham, J.-K. (2008). "Diverse economies: performative practices for other worlds". *Progress in Human Geography* **32**(5): 613–632.

Graham, S. and S. Marvin (2001). *Splintering Urbanism: Networked Infrastructures, Technological Mobilities and the Urban Condition*. London, Routledge.

Hardie, I. and D. Mackenzie (2007). "Assembling an economic actor: the *agencement* of a hedge fund". *The Sociological Review* **55**(1): 57–80.

Hinchliffe, S. (1996). "Technology, power and space – the means and ends of geographies of technology". *Environment and Planning D: Society and Space* **14**(6): 659–682.

Holloway, J. (1998). "'Undercurrent affairs': radical environmentalism and alternative news". *Environment and Planning A* **30**(7): 1197–1217.

Hughes, T. P. (1983). *Networks of Power: Electrification in Western Society 1880–1930*. Baltimore, MD, The Johns Hopkins University Press.

Jasanoff, S., G. E. Markle, J. C. Petersen and T. J. Pinch (eds) (1995). *Handbook of Science and Technology Studies*. London, Sage.

Jones, A. (2013). "Geographies of production I: relationality revisited and the 'practice shift' in economic geography". *Progress in Human Geography*. Online at: http://openaccess.city.ac.uk/2603/1/SR%20PiHG%20Geographies%20of%20Production%20Report%201%2024%20Jul13%20FINAL.pdf (accessed 11 August 2014).

Kemp, R. (1994). "Technology and the transition to environmental sustainability". *Futures* **26**(10): 1023–1046.

Latour, B. (1991). "Technology is society made durable". *A Sociology of Monsters: Essays on Power, Technology and Domination*. J. Law (ed.). London and New York, NY, Routledge: 103–131.

Law, J. (1992). "Notes on the theory of the actor network: ordering, strategy and heterogeneity". Online at: www.lancaster.ac.uk/fass/sociology/research/publications/papers/law-notes-on-ant.pdf (accessed 11 August 2014).

Law, J. (2001). "Ordering and obduracy". Online at: www.lancaster.ac.uk/sociology/research/publications/papers/law-ordering-and-obduracy.pdf (accessed 11 August 2014).

Law, J. and J. Hassard (eds) (1999). *Actor Network Theory and After*. Oxford, Blackwell Publishing/The Sociological Review.

Lemke, T. (2002). "Foucault, governmentality, and critique". *Rethinking Marxism* **14**(3): 49–64.

Leyshon, A., R. Lee and C. C. Williams (eds) (2003). *Alternative Economic Spaces*. London, Sage.

MacKenzie, D. (2003). "An equation and its worlds: bricolage, exemplars, disunity and performativity in financial economics". *Social Studies of Science* **33**(6): 831–868.

MacKenzie, D. (2006a). *An Engine, not a Camera: How Financial Models Shape Markets*. Cambridge, MA, MIT Press.

MacKenzie, D. (2006b). "Is economics performative? Option theory and the construction of derivatives markets". *Journal of the History of Economic Thought* **28**(1): 29–55.

MacKenzie, D. (2007). "The political economy of carbon trading". *London Review of Books*

29(7): 29–31. Online at: from www.lrb.co.uk/v29/n07/mack01_.html (accessed 10 April 2007).

MacKenzie, D., F. Muniesa and L. Siu (eds) (2007). *Do Economists Make Markets? On the Performativity of Economics*. Princeton, NJ, Princeton University Press.

Markard, J., R. Raven and B. Truffer (2012). "Sustainability transitions: an emerging field of research and its prospects". *Research Policy* **41**(6): 955–967.

Mitchell, T. (2008). "Rethinking economy". *Geoforum* **39**: 1116–1121.

Mitchell, T. (2011). *Carbon Democracy: Political Power in the Age of Oil*, New York, NY, Verso Books.

Moss, T. (2001). "Battle of the systems? Changing styles of water recycling in Berlin". *Infrastructures in Transition: Urban Networks, Buildings, Plans*. S. Guy, S. Marvin and T. Moss (eds). London, Earthscan: 43–56.

Muniesa, F., Y. Millo and M. Callon (2007). "An introduction to market devices". *Market Devices*. M. Callon, Y. Millo and F. Muniesa (eds). Oxford, Blackwell Publishing/The Sociological Review: 1–12.

Munro, M. and S. J. Smith (2008). "Calculated affection? Charting the complex economy of home purchase". *Housing Studies* **23**(2): 349–367.

Murdoch, J. (1997). "In human/nonhuman/human: actor-network theory and the prospects for a nondualistic and symmetrical perspective on nature and society". *Environment and Planning D: Society and Space* **15**(6): 731–756.

Murdoch, J. (1998). "The spaces of actor-network theory". *Geoforum* **29**(4): 357–374.

Murdoch, J. (2001). "Ecologising sociology: actor-network theory, co-construction and the problem of human exemptionalism". *Sociology* **35**(1): 111–133.

Oels, A. (2005). "Rendering climate change governable: from biopower to advanced liberal government". *Journal of Environmental Policy and Planning* **7**(3): 185–207.

Pinch, T. and R. Swedberg (eds) (2008). *Living in a Material World: Economic Sociology Meets Science and Technology Studies*. Cambridge, MA, MIT Press.

Rabinow, P. (ed.) (1984). *The Foucault Reader: An Introduction to Foucault's Thought*. St lves, Clays Ltd.

Schot, J. W. (1992). "Constructive technology assessment and technology dynamics: the case of clean technology". *Science, Technology and Human Values* **17**(1): 36–56.

Schot, J. W., R. Hoogma and B. Elzen (1994). "Strategies for shifting technological systems: the case of the automobile system". *Futures* **26**(10): 1060–1076.

Smith, S., M. Munro and H. Christie (2006). "Performing (housing) markets". *Urban Studies* **43**(1): 81–98.

Smith, A., A. Stirling and F. Berkhout (2005). "The governance of sustainable socio-technical transitions". *Research Policy* **34**: 1491–1510.

Smith, A., J.-P. Voß and J. Grin (2010). "Innovation studies and sustainability transitions: the allure of the multi-level perspective and its challenges". *Research Policy* **39**(4): 435–448.

Unruh, G. C. (2002). "Escaping carbon lock-in". *Energy Policy* **30**(4): 317–326.

Velthius, O. (2005). *Talking Prices*. Princeton, NJ, Princeton University Press.

Weber, K. M. (2003). "Transforming large socio-technical systems towards sustainability: on the role of users and future visions for the uptake of city logistics and Combined Heat and Power generation". *Innovation: The European Journal of Social Science Research* **16**(2): 155–175.

2

MAKING CARBON OFFSETS[1]

Key findings

- Two purposefully formed markets (compliance and voluntary) designed to mitigate climate change create the same type of commodity – a carbon offset – but in different ways.
- Because the product of these markets – a carbon offset – is both new and intangible, a great amount of attention has focused on establishing its credibility. The voluntary and compliance markets have different ideas about credibility, and different strategies and approaches designed to achieve it.
- Although at first glance it appears that these two markets have been made from scratch, they have their origins in pre-existing practices, ideas and expertise. There is a path dependency to the pattern of change.
- Polarised debate on offsetting (good or bad) tends to overlook the high degree of variation in carbon offsets, arising from how and where they are produced. The concept of heterogeneous networks draws our attention to this detail of how each carbon offset is made, revealing a multitude of modes of production, involving a range of technologies.
- The objective of the compliance offset market is to actively dissociate the offset from its origins, so that carbon offsets are fully fungible (i.e. tradable and interchangeable), with the aim of making how and where the offset was produced irrelevant. The objective of the voluntary offset market is the opposite: the origins of the offset are actively promoted via particular stories and images about its production.
- There is evidence that the more flexible voluntary carbon market has acted as a 'test bed' for new carbon offset technologies and practices and thus forms a type of innovation niche (a concept from transitions theory about protected spaces where radical innovations are possible).

Introduction

I was fortunate enough to be undertaking fieldwork at a voluntary carbon offset organisation during a pivotal moment in its negotiations with a large American investment bank concerning a possible takeover bid. I was there for a week in 2007, conducting interviews and carrying out an organisational form of participant observation (see further discussion of the methodology below). It was agreed that I could sit in on meetings. The meeting room was packed one morning with several men engaged in what looked to be intense discussions. "Could I sit in on that meeting?" I asked. The shifty looks made me immediately aware that I had asked an inappropriate question. "No, that is not possible," came the eventual reply, "We can't even tell you what that meeting is about". I later learnt, when it became public knowledge, that the meetings that took place over the course of that week led to the purchase of this organisation by the American bank JP Morgan Chase for an undisclosed sum. What I observed through the glass windows of that small meeting room was a key moment of change for the voluntary offset market; and one that brought low carbon economies much closer to the organisations and culture of 'mainstream' global finance.

Whilst carbon offsetting has of late rather fallen out of favour – largely because of the stagnation in international climate negotiations and the global recession (CDM Policy Dialogue 2012; Kossoy and Guigon 2012) – the carbon markets that produce offsets are nonetheless an interesting and important area of study. This is because they are a prime example of the deliberate making of a new market in order to mitigate climate change, centred on the commodity of 'carbon' ('carbon' is here taken as shorthand for any type of greenhouse gas emission reductions).[2] Carbon offsetting at its simplest allows carbon to be reduced in the global atmosphere by compensating for excess emissions in one location through carbon reductions in another (Bumpus and Liverman 2008). The focus of this chapter is how carbon offsets are produced: how the complex mix of offset technologies, techniques, practices and rules come together, and with what results. I concentrate in particular on differences between the voluntary and compliance offset markets, suggesting that these two markets have distinct ideas about how best to achieve their objectives and establish credibility for their products: carbon offsets. The theme of heterogeneous networks is employed here for the overall conceptual analysis within Chapter 2, because understanding carbon markets involves considering not just the social world of policymakers, offset organisations and project developers, but also the material world, the technological substance of these markets: greenhouse gas meters, solar panels and so on.

It is noted in the introduction to Part I that scholars of governmentality and STS use the term 'technology' in different ways. The Foucauldian governmentality definition of technology centres on practices and 'techniques and devices' (e.g.

audits and systems of measurement), rather than engineered material artefacts (i.e. the more conventional definition, referring, for example, to cooking stoves or wind turbines). This definitional difference hints at a more substantial issue, returned to in the conclusion of this chapter. To avoid confusion, and to comply with the more common usage of the terms, the term 'techniques' will be used here to describe the governmentality conception of technologies, and 'technologies' will be used to describe engineered material things – the STS definition. The advantage of attending to governmentality techniques as well as STS technologies is that it provides a set of conceptual tools that covers the spectrum of activities that offsetting entails. These activities encompass the carbon offset supply chain, including the extensive audit and calculation techniques that comprise the governance system for offsetting (the Clean Development Mechanism (CDM) in particular), through to the engineered, material technologies on the ground that comprise offset projects.

The chapter is structured as follows. First, the empirical research and methodology underlying the analysis in this chapter is briefly summarised. Second, carbon markets and carbon offsetting are explained in more detail. Third, voluntary and compliance market approaches to governing carbon offsets are compared and contrasted, focusing particularly on how technologies and techniques are variously used to control and standardise the production of carbon credits and to ensure "the conduct of conduct" (a term used by Foucault to define government, implying self-regulation and direction; see Dean 1999: 11). Fourth, transitions theory is used to explore a key empirical finding – namely, that the voluntary offset market has acted as an experimental 'test bed' for the compliance market. In conclusion, the practical and conceptual implications of having two distinct types of carbon market in operation – compliance and voluntary – are explored.

Empirical research and methodology

The empirical research that forms the basis of this chapter was undertaken during a period of growth in, and excitement about, carbon markets and carbon offsetting – namely, 2007 to 2008. It thus captures a particular moment in the evolution of market-based approaches to mitigating climate change; a time when overall confidence in both the voluntary and compliance markets was strong. It was a period of rapid transition and hence uncertainty, but also of optimism amongst those involved in climate mitigation (both within practitioner and policy circles) about the potential for purpose-built markets to mitigate climate change. Carbon offset markets were in the process of being made, and being made fast. The empirical research was undertaken as part of a team of UK Tyndall Centre researchers, led by Professor Diana Liverman (then at Oxford University) (see Tyndall Centre 2008). The focus was on the 'upstream' offset organisations involved in making carbon offsets. The specialist knowledge, technologies and discourses associated with offsetting mean that offset organisations are key players in understanding how carbon offsets are produced and consumed, and so we concentrated on these offset organisations in this strand of our Tyndall Centre research.

During 2007 and early 2008, I conducted twenty-five interviews with offset organisations, several jointly with Professor Diana Liverman. The interviews were mostly with managers and directors of two particular international organisations (one operating in the compliance and one in the voluntary offset market). We selected these two organisations for our research, because at the time they were large, well-known and had engaged with pioneering projects, methods and markets: they were 'market leaders' with considerable presence in policy discussions and international meetings. They were also based in the UK and were willing to talk to us and allow us to track their activities over the period of our empirical research.

In addition to the expert interviews, the empirical research involved: analysis of the grey literature on offsetting, including company reports and websites, news reports, evaluations by non-governmental organisations (NGOs) and others; participant observation of the interactions that offset organisations had with other actors at the Bali 2007 and Poznan 2008 United Nations Framework Convention on Climate Change (UNFCCC) Conference of Parties (the annual international meeting of signatories to the UNFCCC),[3] as well as at several public workshops on offsetting during this period. The objective of Chapter 2 is to capture and better understand an especially active, critical period in the making of voluntary and compliance markets (2007–2008). It does not, therefore, provide a comprehensive, up-to-date review of carbon markets (see instead, CDM Policy Dialogue 2012; Kossoy and Guigon 2012).

Explaining carbon offsetting

As noted, there are two broad types of carbon offset market: compliance (primarily the Clean Development Mechanism under the Kyoto Protocol) and voluntary. The CDM is one of the three United Nations Kyoto Protocol 'flexible mechanisms' (the others are Joint Implementation and International Emissions Trading; see Capoor and Ambrossi 2009). The CDM allows developed countries to comply with their target emission reductions through purchasing emission reductions made elsewhere, specifically within developing countries with no targets under the Kyoto Protocol, rather than through domestic reductions. The demand for carbon reductions is driven by the commitments made by industrialised countries that ratified the Kyoto Protocol and thereby agreed to reduce emissions from a 1990 baseline by an average of 5.2 per cent by 2012. Emission reduction commitments post-2012 are still in the process of being agreed, with a new agreement anticipated at the UN annual climate change meeting to be held in late 2015 (UNFCCC 2014).

Creating the demand and supply necessary for a market in something as intangible as greenhouse gas emission rights and obligations implies numerous acts of quantification, measurement and commensuration (see Ascui and Lovell 2011; and also Part III of this volume). On the one hand, demand was created by placing caps on national emissions from developed countries. On the other hand, supply was created within the CDM by creating an entirely new commodity in the form of an emission right based on an emission reduction achieved in a country without a cap. With the introduction of the Kyoto Protocol's CDM, entirely new carbon accounting rules were therefore required within developing countries to enable the measurement of emission reductions against a hypothetical baseline in defined CDM projects. Such emission reductions gave rise to credits known as Certified Emission Reductions (CERs); a developed country may obtain such CERs and use them to exceed its cap by one tonne of carbon dioxide (or its equivalent in other greenhouse gases) per CER. Thus, the CDM is engaged in an entirely novel project of "making things the same" (MacKenzie 2009: 440): in this case, making reductions in emissions against a baseline equivalent to emission rights in developed countries.

Many companies and individuals also wish to offset their emissions in the parallel voluntary offset market to directly compensate for their greenhouse gas emissions. The voluntary offset market has developed independently of the UNFCCC and is much smaller and more flexible. The main purchasers of voluntary and compliance offsets therefore differ considerably, with corporations buying the majority of voluntary offsets, whereas compliance offsets were formerly mostly bought and sold by governments (Lovell *et al.* 2009), but from 2006 onwards the majority of demand has come from the European Emissions Trading Scheme (see Capoor and Ambrossi 2007). There are a number of voluntary carbon offset standards, which most offset organisations now choose to use for their products, foremost amongst them being the Verified Carbon Standard, with an overall 55 per cent market share (Peters-Stanley and Yin 2013). At the time of undertaking the Tyndall Centre research (2007–2008), however, the situation was quite different: there was a great deal of flux with several new voluntary carbon offset standards emerging and little data on who was using which standard (Hamilton *et al.* 2007). The voluntary market was much more fragmented and informal than it is now. The compliance offset market, in contrast, has always been closely regulated by a complex array of rules at the international level, administered by the CDM Executive Board of the UNFCCC. The CDM has mechanisms to define credits strictly and establish standards of quality for the design and methodology of projects (including those for additionality and baselines).[4] All CDM projects must be registered through the UNFCCC, and carbon finance is typically channelled through private sector companies or World Bank carbon funds, which then finance offset projects in the developing world (Bumpus and Liverman 2008). In order for a technology to be approved for credit generation under the CDM, it must go through a lengthy process of review and assessment, which can take up to a year or more (IETA 2008).

So there are notable differences between these two markets – compliance and voluntary – despite them producing a superficially similar product: a carbon offset. The compliance market is hierarchical and highly regulated, whilst the voluntary offset market remains much more informal, with no overarching definition for credits, varying carbon calculators and several competing voluntary standards. Often described as a 'parallel market', voluntary offset projects tend to be smaller, have a greater sustainable development focus (often described as social or community 'co-benefits'), lower transaction costs, involve a wider range of methods or techniques and are typically located in countries not active in the CDM (e.g. the non-Kyoto signatory the USA and several African countries) (House of Commons Environmental Audit Committee 2007). Unlike offset organisations working within the structure imposed by the UNFCCC and the CDM, voluntary offset organisations can use flexible practices through more informal networks with NGOs, or companies active in the Global South, to source projects and ultimately generate credits (Taiyab 2006).

One common source of misunderstanding about carbon offsets is a lack of appreciation of this diversity across carbon offset markets. Diversity is apparent not only in the forms of governance (compliance and voluntary), but also in terms of the technologies that can be used to produce an offset. Critics have often overgeneralised following the failure of a handful of projects in the voluntary market (FERN 2005; SinksWatch 2007) or from the use of the CDM to fund greenhouse gas reductions at a few industrial facilities in China (Wara 2007). Table 2.1 shows the carbon offset projects approved by the CDM Executive Board to date and the number of credits issued per technology type. Note that these are approved technologies for compliance projects only. Data on voluntary offset technologies reveals a somewhat different picture: with forest offsets (afforestation/reforestation and REDD) forming 21 per cent of market share of voluntary offsets, wind (20 per cent) and cooking stoves (8 per cent) (Peters-Stanley and Yin 2013: x). As the data in Table 2.1 demonstrates, the most common types of offset project under the CDM are hydro power, biomass and wind: these three technologies comprise over half of all registered CDM projects. If considering the number of carbon credits actually issued, however, then industrial gas projects (typically involving adding scrubbers to remove greenhouse gases such as hydrofluorocarbons (HFCs) and nitrous oxide (N_2O) from exhaust fumes) are most significant, representing 55 per cent of credits.

Table 2.1 Technologies used to produce carbon offsets in the Clean Development Mechanism

Type of CDM projects by technology	*CDM projects with CERs issued*			
	No. of projects	*%*	*Issued CERs (1,000s)*	*%*
Hydro	754	30	181,207	13
Wind	726	28	144,385	10
Biomass energy	240	9	41,214	3
Methane avoidance	209	8	19,601	1
Energy Efficiency – own generation	152	6	67,403	5
Landfill gas	139	5	53,692	4
N_2O	57	2	267,802	19
Fossil fuel switch	49	2	49,587	3
Coal bed/mine methane	38	1	31,719	2
Energy Efficiency – industry	38	1	2,911	0
Solar	26	1	1,134	0
HFCs	19	1	519,134	36
Reforestation	18	1	10,172	1
Energy Efficiency – supply side	13	1	3,117	0.2
Cement	12	0.5	3,024	0.2
Energy Efficiency – households	12	0.5	424	0.0
Fugitive	12	0.5	21,151	1
Geothermal	10	0.4	7,193	1
PFCs and SF_6	9	0.4	4,481	0.3
Transport	7	0.3	1,141	0.1
Energy distribution	5	0.2	1,501	0.1
Afforestation	3	0.1	751	0.1
Mixed renewables	2	0.1	16	0.0
CO_2 usage	1	0.0	10	0.0
Energy Efficiency – service	1	0.0	9	0.0
Tidal	1	0.0	433	0.0
Total	2,553		1,433,215	

Source: UNEP Risoe CDM/JI Pipeline Analysis and Database, 1 April 2014. Online at: www.cdmpipeline.org/ (last updated 1 March 2014).

Table 2.2 shows a further breakdown of CDM project sub-types under a selection of the main technology categories listed in Table 2.1, in order to illustrate further the complex diversity of technologies that comprise compliance offset projects. Thus, for example, the category of 'biomass energy' comprises over ten possible sub-types of project, ranging from electricity generation from biomass residues to use of charcoal in iron ore reduction. Each of these biomass technologies falls under a separate CDM methodology, which varies in terms of the technology used, the production process and the typical volumes of carbon credits produced. CDM projects also vary in terms of their impact on local communities and associated sustainable development and environmental co-benefits. Nonetheless, these CDM projects all generate identical carbon credits – Certified

Table 2.2 Further breakdown of selected Clean Development Mechanism project technologies by sub-type

Type	*Sub-types of project in each CDM technology category*
Biomass Energy (585 projects)	Emission reduction through partial substitution of fossil fuels with alternative fuels in cement manufacture (44) Grid-connected electricity from biomass residues (353) Electricity generation from biomass residues in power-only plants (81) Co-firing of biomass residues for heat generation and/or electricity generation in grid-connected power plants (0) Analysis of the least-cost fuel option for seasonally operating biomass co-generation plants (0) Fuel switch from fossil fuels to biomass residues in boilers for heat generation (24) Grid-connected electricity generation using biomass from newly developed dedicated plantations (2) Use of charcoal from planted renewable biomass in the iron ore reduction process through the establishment of a new iron ore reduction system (4) Switch from non-renewable biomass for thermal applications by the user (27) Energy efficiency measures in thermal applications of non-renewable biomass (45) Switch from fossil fuel to biomass in existing manufacturing facilities for non-energy applications (5)
N_2O (120 projects)	Decomposition of N_2O from existing adipic acid production plants (4) Catalytic N_2O destruction in the tail gas of nitric acid or caprolactam production plants (27) Catalytic reduction of N_2O inside the ammonia burner of nitric acid plants (62) Secondary catalytic N_2O destruction in nitric acid plants (0) N_2O abatement from nitric acid production (27)
Energy efficiency – supply side (174 projects)	Conversion from single cycle to combined cycle power generation (29) New grid-connected fossil fuel fired power plants using a less GHG intensive technology (52) Natural gas-based package co-generation (15) New co-generation facilities supplying electricity and/or steam to multiple customers and displacing grid/off-grid steam and electricity generation with more carbon-intensive fuels (8) Increased electricity generation from existing hydropower stations through decision support system optimisation (3) Rehabilitation and/or energy efficiency improvement in existing power plants (3) Energy efficiency improvement of a power plant through retrofitting turbines (4) Implementation of fossil fuel trigeneration systems in existing industrial facilities (1) Installation of co-generation system supplying electricity and chilled water to new and existing consumers (2) Installation of a new natural gas fired gas turbine to an existing CHP plant (3) Greenfield co-generation facility supplying electricity and steam to a Greenfield industrial consumer and exporting excess electricity to a grid and/or project customer(s) (2) New natural gas based co-generation plant (10) Introduction of an efficiency improvement technology in a boiler (0) Supply side energy efficiency improvements – generation (31) Recovery and utilisation of waste gas in refinery facilities (8) Electricity and/or heat generation using fuel cell (0) Conversion from single cycle to combined cycle power generation (3)

Source: UNEP Risoe CDM/JI Pipeline Analysis and Database, 1 April 2014. Online at: www.cdmpipeline.org/ (last updated 1 March 2014).

Emission Reductions (CERs), equal to one tonne of carbon dioxide (or its equivalent in other greenhouse gases) – that, in theory at least, are fully fungible and interchangeable. Indeed, the system of compliance offsetting hinges on the crucial dissociation between the offset production process and the resulting carbon credit. In other words, it is key to the functioning of the compliance offset market that the technical diversity of offsetting – the project-to-project variation in how the offset is produced – is disregarded. Over time, however, achieving this fungibility in compliance carbon offsets has become harder and harder (see CDM Policy Dialogue 2012). Although at the time of undertaking the Tyndall Centre research, there was evidence of some credits produced by particular technologies reaching a lower market value (e.g. credits from industrial gas projects and forests; see Carbon Finance 2007b), in broad terms the aim of credit fungibility had been achieved: a Certified Emission Reduction (CER – a CDM offset) had a uniform price in 2007–2008 and could be freely traded on a like-for-like basis without reference to its origins. In order to achieve this ambitious policy goal, a substantial amount of work had to be put into devising a system of calculation, measurement and audit aimed at "making things the same" (MacKenzie 2009: 440).

In contrast, within the voluntary offset market carbon credits have never been fully fungible – the market was never designed to achieve this. The voluntary market has always had a much less extensive governance system to measure, verify and audit voluntary carbon credits, termed 'VERs' (Verified or Voluntary Emission Reductions) (Bumpus and Liverman 2008; Kollmuss *et al.* 2008; Lovell *et al.* 2009). In the voluntary market, the focus is much more on the communities and people involved in offset projects, not just the greenhouse gas emission reductions. As is argued elsewhere (Lovell *et al.* 2009), the discourse about carbon offset projects is particularly important in the voluntary offset market: certain carbon credits are attractive because they have a story associated with them and can be sold at a premium as 'gourmet' or 'boutique' carbon, with an emphasis on their poverty-alleviation 'co-benefits'. In other words, information and knowledge about how the offset is produced – where and using what technology – is crucial within the voluntary offset market. This is in stark contrast to the compliance market, where this type of information is actively dissociated from the credit (or at least attempted to be dissociated, as discussed). This is a key distinction between the two markets and demonstrates how markets can be made in quite different ways, even when, as here, they have been purposefully created to provide mitigation of the same problem: human-induced climate change.

Another, related distinction between the voluntary and compliance offset markets is the manner in which credibility is defined and established: how an intangible, novel commodity – the carbon offset – is made legitimate. In the compliance market, credibility stems primarily from the rigorous, extensive audit practices to do with sourcing greenhouse gas reduction projects, managing their assessment and verification and issuing the resulting credits for sale. Practitioners in carbon markets call these procedures 'measurement, reporting and verification', abbreviated to 'MRV' (see also Chapter 6). MRV is also an important part of voluntary markets, but it plays a role alongside other techniques: credibility for the voluntary offset product stems additionally from the narratives linking voluntary offset consumers and producers, in a similar way to fair trade and organic products (Lovell *et al.* 2009). Applying ideas about heterogeneous networks – such as being attentive to these differences in how carbon offsets are produced and sold and exploring the relations between the mix of material and social things that comprise carbon offset markets – provides a means to better understand the complexity of these purpose-built markets. This chapter now turns to consider these conceptual issues and opportunities in more detail.

Theorising the role of technology in offsetting

Here, ideas about heterogeneous networks are used to explore the agency of technology and its important role in shaping social relations, politics and governance processes within carbon offset markets. As outlined in the introduction to Part I, the central proposition of a heterogeneous networks approach to the study of markets and economies is that it is not just social actors that are important, but material ones, too. Through a range of empirical examples, it is demonstrated how the technologies and techniques of carbon offsetting are multiple, complex and diverse. A question is raised regarding the interplay between, and different meanings of, technologies and techniques within carbon markets. For, as noted, within the CDM the MRV techniques that underpin this carbon market – the processes of measurement, calculation, audits and so on – are at least as important to making this market as are the greenhouse gas reduction 'technologies' (in the material and common sense use of the term, those technologies that are producing greenhouse gas emission reductions within a project – e.g. a hydro power plant or industrial gas scrubber).

Achieving credibility in heterogeneous networks

Starting the analysis with the compliance offset market, the concept of heterogeneous networks provides a framework for understanding the complex audit process of the CDM and its obsession with calculation: in project design, verification, the CDM Executive Board decision-making process and so on. The CDM represents an elaborate attempt to control and order greenhouse gas emission reductions, where the building and maintenance of a credible system of governance has become the main focus of policy and professional effort. The primary focus of the day-to-day governance of climate mitigation has therefore been on making the audit and verification system work, at some distance from the fundamental objective of reducing net emissions of greenhouse gases into the atmosphere. The CDM governance process has been heavily criticised for becoming too cumbersome because of its obsession with accuracy, as one offset company director with fifteen years' experience in carbon markets explains:

> [T]he CDM Executive Board is very, very concerned with getting things exactly right, as opposed to being conservative and recognising that you don't know every piece of data on every single project. That has been a key concern . . . the current CDM insists on extreme exactitude around every single project and every single asset. . . . It is realising the absurdity of that . . . there has to be an adaptation of the approach.
>
> (Interview, director of a carbon offset organisation, January 2008)

Conceptual insights from governmentality complement more STS-orientated ideas about heterogeneous networks by highlighting how the credibility of the CDM has become linked to its 'techniques' of government – namely, processes of measurement, calculation and standardisation. For instance, Li (2007), in her insightful analysis of development projects in Indonesia, views calculation and defining objects of government in a technical way as vital to modern practices of governmentality, as she explains:

> Calculation is central, because government requires that the 'right manner' be defined, distinct 'finalities' prioritized, and tactics finely tuned to achieve optimal results. Calculation requires, in turn, that the processes to be governed be characterized in technical terms. Only then can specific interventions be devised.
>
> (Li 2007: 6)

Although the empirical focus of Li's work is somewhat different, her ideas can readily be applied to the CDM. Within the CDM, the problem of climate mitigation has been progressively more narrowly framed, in ways that can only be addressed and managed by international systems of CDM audit and calculation. As other commentators have noted, there is significant and constant behind-the-scenes effort in carbon markets to 'make things the same' (Lohmann 2005; MacKenzie 2009), to maintain the fungibility of carbon credits, despite their production involving a large diversity of offset technologies in many different places. In the period from 2007 to 2008 – at the time of undertaking our research on carbon offsets – this objective had mostly been achieved, but, as noted above, it has more recently run into difficulties, stemming largely from the knock-on effect of the failure of the Copenhagen COP climate talks in 2009 (CDM Policy Dialogue 2012). But the tensions in making things the same within carbon markets have always been present, even if at the outset they remained mostly hidden or 'black boxed'. Bumpus (2009), for example, highlights a number of tensions in the production of carbon offsets in his in-depth analysis in the period 2006–2007 of the practices and materiality of carbon offset projects in Honduras. He provides several examples of how carbon is an unusual and 'slippery' commodity, in part because of the uncertain nature of carbon offset technologies: in the La Esperanza micro-hydro CDM project, for instance, the generator blew up because of unexpectedly heavy rainfall. Similarly, Boushel (2014), in her in-depth study of a carbon offset standard organisation, provides many instances of struggles to maintain the fungibility and credibility of carbon offsets as a new type of product, from the project level and beyond. We need to recognise the manifold possibilities for technical complications at a project level to understand how, over time, the focus on calculation in the CDM has led to the prioritisation of low carbon technologies that produce greenhouse gas emission reductions that can be relatively easily measured and controlled (Lohmann 2009). The audit process – compliance offset MRV – is itself shaping the market and hence stretching beyond its intended background role, as others have similarly found with audits in other types of markets (Power 1999). For example, a manager at an offset organisation describes their efforts to bring less easily quantifiable greenhouse gas reduction technologies into the CDM framework:

> What we are struggling with right now is how can we innovate in some sectors . . . the cool, co-benefits kind of projects that buyers are wanting . . . but that are very difficult to implement from a CDM or traditional method of doing carbon credits. You just cannot do entire PDDs [Project Design Documents] and come up with huge baseline analyses and do the very complicated monitoring up to the tenth decimal point for a project that is going to do small scale biodigesters in Nepal. . . . In order to make it work in a sensible way you need to have new protocols and new ways of thinking that are outside of the box.
>
> (Interview, manager at an offset organisation, April 2008)

Interestingly, here, the interviewee (a manager at a large compliance offset organisation, who had just moved into a new role involving voluntary offsets) is implying that it is the voluntary market that represents an opportunity for developing innovation, a point returned to below. It indicates that those working in compliance carbon markets are acting in a manner that is quite limited and constrained in terms of the types of low carbon technologies and techniques they are able to use. The heterogeneous network is tightly framed (see Part II), providing advantages for market participants who benefit from 'knowing' the market (Callon 1998) and also constraints that perhaps do not ultimately assist in mitigating climate change. Again, Li provides insight into this process of framing and technical constraint: "A central feature of [development project] programming is the requirement to frame problems in terms amenable to technical solutions. Programmers must screen out refractory processes to circumscribe an arena of intervention in which calculations can be applied" (Li 2007: 2).

Techniques such as audits are often borrowed or adapted from other areas and applications. An explicit programme of policy intervention is a rare thing. More often than not, it is a heterogeneous network – made using a blend of past practices and ideas. Indeed, a prime example of this is the migration of techniques (albeit, a partial, sometimes contested migration) from the financial sector to carbon markets, as an offset organisation director explains:

> The bank traders want something that will fit into their existing systems. The trader sits there with two lines and if it goes over the top of that line, they sell it. When it goes under the bottom of that line they buy . . . all they have got to do is watch the screen. So from the point of view of a three dimensional instrument [like a carbon offset] whereby actually there's all these other factors, it's not just price and volume, they actually don't have the mechanisms.
>
> (Interview, CEO of an offset organisation, October 2007)

So although carbon offset markets are new, they incorporate and hybridise other existing ideas, practices and technologies as they emerge and evolve. In other words, they are not made from scratch, but rather have grown out of other, already existing markets and economies. Compliance carbon offset markets have been forged in a rather different arena to voluntary offset markets, one of international governance (the UNFCCC) and global finance. Analysis now turns to consider the contrasting origins and characteristics of the voluntary offset market.

As discussed, the voluntary market has concentrated much less on auditing and the regulatory system and more on individual offset projects and the human-technology relations within them.[5] In essence, the focus has been more 'bottom-up' than 'top-down': on specific greenhouse gas reduction technologies – how they work and how socially integrated they are in specific localities – rather than governance techniques and practices. In this way, the voluntary market shows just how different climate mitigation markets can be.

Voluntary offset markets are heterogeneous networks that are made locally. A manager at an offset organisation contrasts the voluntary sector focus on the project with the compliance market as follows:

> Obviously, the compliance market doesn't care where it [the offset] comes from. The compliance market just wants to make sure it's verified and signed off and, boom, off you go. Whereas in the voluntary market it's very, very dependent on what the technology is – very dependent. Our pricing structure reflects that.
>
> (Interview, voluntary market manager at an offset organisation, April 2008)

Thus, particular climate mitigation technologies such as cookstoves and small-scale hydro plants are able to sell their offsets at a premium within the voluntary market, as so-called 'boutique' or 'gourmet' carbon (Bumpus 2011). Their value is higher because embedded within the product, as a significant part of what is being sold, is a narrative about the integration of the technology into everyday lives in developing countries (Lovell *et al.* 2009). A core aim of the voluntary market is developing this strong, direct connection between producers and consumers, as a voluntary offset manager explains:

> . . . people want to feel a connection where the money is being spent and that is much, much easier for people if we talk about communities and families and school and day-to-day [things] and not big fat technical processes . . . it will be harder to sell the idea of parting with some of your cash for carbon reductions if you don't have those sort of stories.
>
> (Interview, marketing manager at an offset organisation, October 2007)

Material technology, used to effect reduction in greenhouse gases, and the social connections of that technology are both integral to the making of voluntary offset markets. This integration is in keeping with heterogeneous network approaches to markets and economies, which ascribe a key role to materials or technologies (the non-human elements of policy) in processes of change (Latour 1992). Further, by translating an idea into a material form – whether it be constructing a new type of power plant or demonstrating a new renewable energy technology – the idea is given some permanence. Material technologies thereby play a particularly important role in adding stability to new heterogeneous networks such as carbon offsets, as Law explains: ". . . some materials are more durable than others and so maintain their relational patterns for longer . . . Consequently, a relatively stable network is one embodied in and performed by a range of durable materials" (Law 1992: 5).

The key insight here for carbon markets – both voluntary and compliance – relates to their newness and the consequent need for stability. Carbon offsets are a new commodity with heterogeneous components and comprise a complex mix of part public, part private and part non-governmental/third sector entities.

Because carbon markets have remained 'hot' or unsettled (Lohmann 2005; Callon 2009), and the institutions and practices of carbon offsetting are only relatively weakly embedded, technologies can be seen to play a formative role in providing stability and credibility and in connecting diverse social actors. It is evident that during the period of study (2007–2008) relationships between technologies and humans within these heterogeneous networks were still actively in the process of being worked out, as the director of a voluntary offset company explains:

> Two years ago there were two of us in this [voluntary offset] market and now there are 120. The rate of growth . . . the intervention of clumsy government, the uncertainty . . . means that it's just going be turbulent. It's not like selling widgets that have been sold for the last 35 years . . . we're on a river, the river's going down, eventually we know it's going to reach the sea but there's going be some waterfalls on the way . . .
>
> (Interview, director of an offset organisation, October 2007)

Indeed, as this interviewee accurately predicts, there have been a number of 'waterfalls' in the path of carbon offset markets since 2007, including, for example, a substantial drop in price and volume of carbon credits, as well as continued uncertainty about the future of the compliance offset market as international negotiations have stalled (CDM Policy Dialogue 2012).

Sociotechnical transitions and carbon offset markets

As noted in the introduction to Part I, whilst heterogeneous network approaches such as *agencement* and actor-network theory are good at focusing our attention on the relations within relatively discrete, static markets and economies, they provide fewer insights regarding larger-scale, radical change over time. In this regard, transitions theory is a useful complement, because at its core is a recognition of the path dependency of technologies: acknowledging that technologies themselves have a history; they have been used in different places and for different purposes in the past, and this can, in turn, influence their contemporary role and social meaning (Arthur 1989; Schot *et al.* 1994; Unruh 2002). To take cookstoves as an example: more efficient stoves, as a replacement for traditional three-stone fires, were used for decades prior to the emergence of carbon offsetting and have been extensively funded under previous programmes of development and poverty alleviation (see NC State 2014). Cookstoves have now been actively 'reframed' in the voluntary offset market as being connected with carbon, rather than development. This reframing is the innovation: they are new as offsets, but obviously not new as technologies. It is a *socio*technical innovation. Attention to the path dependency of specific technologies can further our understanding and appreciation of why technologies, reframed as carbon offsets, might be resisted or embraced in particular locales, depending on their history of use. Already existing technologies and materials – in use before the birth of carbon markets – can act as a 'script' (Akrich 1992), placing boundaries on the acceptable limits of new policy initiatives.

The idea of path dependency is also helpful when applied not just to technologies, but also to broader Foucauldian rationalities (ideas, practices and types of knowledge). In this way, differences between the compliance and voluntary offset markets can be attributed not just to the type of technologies they use (and the path dependency of each technology), but also to their different institutional origins. Each market has borrowed ideas, techniques, expertise and practices from nearby: for the compliance market, its original 'home' is intergovernmental organisations (especially the United Nations) and banking; for the voluntary offset market, it is more the arena of international development NGOs, green entrepreneurialism and fair trade/organic products (Lovell *et al.* 2009). I now explore the interactions between the voluntary and compliance offset markets, drawing on transitions theory, to better understand the pathways and patterns of change in these two markets.

The voluntary offset market as an innovation niche

Analysis of interviews conducted with managers of offset organisations reveals evidence of the voluntary offset market providing a type of experimental space or 'test bed' for low carbon ideas and technologies. These technologies are yet to be adopted within the more regulated, formalised compliance offset market. Unlike offset organisations working within the rules-based structure imposed by the UNFCCC and the CDM, voluntary offset organisations can use flexible practices through more informal networks to source projects and generate credits (Taiyab 2006). A key finding from our Tyndall Centre research was that it was principally from within the voluntary offset market that new ideas and approaches to offsetting were emerging. As the sales manager at a voluntary offset organisation explained:

> . . . the voluntary market can act as a testing bed for the compliance market for new ideas and approaches and technologies. The voluntary market has a really big role to play in innovating and trying new ways out and sometimes they will fail but you haven't got the enormous costs of the compliance market.
>
> (Interview, sales manager, voluntary offset organisation, October 2007)

In his interpretation, the voluntary market is conducive to innovation because the costs of experimentation are lower. The founder and president of an international compliance offset organisation explained his reasons for developing new business in the voluntary offset market in somewhat similar terms, as follows: "We were excited about having the voluntary market grow because it would bring back to this sector innovation and flexibility. . . . We are really working on getting in this space to be the major player there" (Interview, July 2007).

These portrayals of the role of the voluntary offset market as a flexible, experimental space bear a strong resemblance to the concept of an innovation niche within transitions theory (see the introduction to Part I), defined as something that ". . . protects against too harsh selection and provides space to grow" (Schot *et al.* 1994: 1061). However, the distinctive, 'bottom-up' market-based form of financial protection identified within the voluntary offset market contrasts with most examples of protection provided in transitions theory. Here, 'protection' is provided predominantly via public sector grants and subsidies (Geels 2004; Smith 2007; Smith and Raven 2012). The voluntary market is seen as having greater flexibility and adaptability in comparison with the compliance market for two main reasons: first, because it is cheaper and quicker to set up offset projects, as there are fewer regulatory requirements; and second, because the projects tend to be smaller and typically initiated by local organisations based in the Global South. This means that there are well-established channels of communication at the project level, and hence strong opportunities for one-off experimentation and learning (see, for example, the small-scale agro-forestry projects managed by Plan Vivo 2007). The flexibility of the voluntary market has thus allowed it to become a space for trying out new low carbon technologies and practices that do not fit within the tightly defined rules and regulations of the compliance offset market, as the carbon finance manager at an NGO further explained:

> The CDM has a risk-averse approach to approval of new methodologies, and there is less scope for innovation there than within other schemes. . . . Some examples are the CFL [Compact Fluorescent Lightbulb] projects, they are not allowed under the CDM at the moment. Another one is Avoided Deforestation. That is being trialled in the voluntary market right now and we would expect to see it in the post-2012 [UNFCCC] framework.
>
> (Interview, July 2007)

A voluntary markets manager at a leading compliance offset organisation similarly captured this notion of the voluntary market as an experimental space:

> I would say that the main advantage of the voluntary market is that it offers the possibility for doing innovative stuff. I think that is really one

> of the key things. By 'innovative stuff' I mean really coming up with new protocols and new methods for doing [offset] projects.
>
> (Interview, April 2008)

The interviewee above foresaw a number of offsetting initiatives (Avoided Deforestation, CFL projects) that have, indeed, been subsequently incorporated ('mainstreamed') into the compliance market. 'Avoided Deforestation' has re-emerged in a slightly different form as a UN initiative: 'Reducing Emissions from Deforestation and Forest Degradation' (REDD) (UN REDD 2014; see Chapter 6). CFL projects are also now permitted within the CDM under new initiatives to encourage small-scale projects and 'bundling' of emission reduction activities (Baker & MacKenzie 2014).

As these interviews illustrate, it is not just experimentation and learning about new technologies that have been important within the voluntary market, but also the development of alternative governance practices and more flexible modes of regulation; in essence, trialling different ways of producing and consuming offsets. This is in keeping with the *socio*technical emphasis of transitions theory. During the Tyndall research period, there was considerable concern amongst key players in the carbon market about the impact of bureaucracy, and slow decision-making within the CDM Executive Board, on the commercial viability of compliance offset organisations (Carbon Finance 2007a). Indeed, it was so-called 'CDM bureaucracy' that motivated a number of compliance offset organisations to start engaging more with the voluntary market at this time (2007–2008), as the president of one such company explained: ". . . the [CDM] bureaucracy is contributing to climate change . . . it is delaying mitigation action . . . so much so that a lot of actors are deciding to follow the voluntary path" (Interview, July 2007). Similarly, a manager of a voluntary offset company elaborated:

> I think we, collectively, as a company, have gone naturally to the conclusion that CDM is great wonderful fantastic, but it needs to be re-thought completely. . . . We just can't have ten people in Bonn [on the CDM Executive Board] looking at projects any more. It is just not going to work. So I think there needs to be a paradigm shift in how projects are done. I think one of the good things about the voluntary market is that it allows for experimentation on that.
>
> (Interview, April 2008)

It is not just technical innovations, but also social and institutional innovations that are being tested within the voluntary market: it is a sociotechnical innovation space. It is an arena in which private and third sector/non-governmental organisations often view themselves as better informed and more knowledgeable than public sector actors. In 2007–2008 there was a feeling amongst offset organisations that individual states and the international regime were no longer fully in touch with what was happening 'on the ground' with carbon offsets and carbon markets – compliance, as well as voluntary – and had therefore, to some extent, lost the ability to govern it effectively. Offset organisations were increasingly positioning themselves as having the experience and expert knowledge to make effective decisions about feasible, workable offset rules and regulations (Lovell 2010). Crucially, the opportunities for trying out new modes of governance were seen as more numerous with the voluntary offset market. The press officer at a leading voluntary offset organisation describes the advantages of their flexible 'self-government' and their collective experience and knowledge, relative to UN policymakers, who are presented as relatively new and inexperienced:

> [T]he advantage of giving the voluntary market space to evolve and encouraging it to regulate itself means that you do get the wisdom and input of a lot more people ... [a project developer] brings 30 years' experience of offset project development ... so if he is involved in feeding that [experience] to the people who do the standards then it's a bit more workable. He will probably in a very short space of time develop a workable or best case scenario [for regulation] rather than a few people in a government department for the UN necessarily coming up with best solutions straight away...
>
> (Interview, press officer of a voluntary offset organisation, October 2007)

By drawing on ideas about sociotechnical transitions – in particular, the innovation niche – the analysis here indicates that the voluntary offset market is acting as an innovation niche 'test bed' for the international climate regime (the CDM compliance market). The markets have been made differently, and therefore continue to operate differently, with advantages and disadvantages of both types of market. Analysis reveals the interplay between the two markets, how and what they have learned from each other, and shows the smaller, more flexible voluntary market to be positioned as the innovative forerunner. There are also empirical insights into how the 'mainstreaming' of innovations – their transition from niches to regimes – actually takes place: mainstreaming in the case of carbon offset markets has been a messy process, wherein compliance market actors have 'dipped in and out' of the voluntary market, in order to test new ideas and technologies. In other words, there has not been a straightforward, linear migration of new carbon offset practices from the voluntary offset niche to the compliance regime.

Summary and conclusions

There remains confusion amongst those working outside carbon offset markets about what exactly a carbon offset is. The conceptual theme of heterogeneous networks provides a relevant theoretical lens for understanding the multitude of technologies, ideas, MRV practices and organisations that comprise carbon offset markets. Carbon offset markets have become notably complex. Within the compliance offset market, there is an obsession with calculation and with the accuracy of methods used to measure the production of carbon credits from individual projects that use a vast array of different greenhouse reduction technologies. In the media, and to a lesser extent in academia, the debate about offsetting has become polarised and overgeneralised. Much of the subtlety of the diversity of technologies and techniques of offsetting has been misinterpreted or overlooked (FERN 2005; FoE 2009). Methods of carbon offsetting tend to be lumped together, but the reality is two distinct markets – compliance and voluntary – comprising different types of offset technologies, institutions and practices. This situation provides us with an unusual and informative case study of how low carbon economies and markets are made. These two distinct routes or pathways for making markets ostensibly do the same thing: take greenhouse gases out of the atmosphere. The compliance market is heavily regulated, with a strong focus on audit, verification and measurement: building a credible system of governance has been the focus of policy and professional effort, because measurement processes and techniques are central to the CDM goal of producing standardised (fungible) uniform credits that are dissociated from their origin. The voluntary market, in contrast, has concentrated rather less overall on the governance of credits (and particularly in the period in question, 2007–2008) and more on individual projects and the human-technology relations within them; so the focus has been on the greenhouse gas reduction technologies themselves – the mechanics of them, how they are integrated in specific localities and how their stories are communicated.

There are a number of policy and conceptual implications that stem from the analysis. In terms of policy implications, the limitations of the CDM taking calculation and measurement so seriously are increasingly being recognised. The CDM governance process is now widely perceived as too cumbersome, because of its obsession with accuracy (Lohmann 2005; FoE 2009; CDM Policy Dialogue 2012). The focus of the CDM on MRV techniques has also led to a lack of attention to the operation of low carbon technologies at a project level and neglect of some of the regions of the world with less capacity and smaller scale opportunities. In the period since 2007–2008, these compliance market issues have been formally recognised and acted upon, including via the CDM Policy Dialogue initiative, established in 2011 with the mandate to produce recommendations on

> how to best position the CDM to respond to future challenges and opportunities and ensure the effectiveness of the mechanism in contributing to future global climate action, based on a wide-ranging assessment of experience, benefits and shortcomings of the CDM and engagement with civil society, policymakers and market participants.
>
> (UNFCCC 2012)

But the voluntary market was also having credibility problems in 2007–2008, because of too little monitoring and regulation of offsetting activities: the opposite reason to the compliance market. Since 2008, the voluntary offset market has also changed significantly in recognition of this. By 2012, 98 per cent of voluntary offsets were accredited by standards (Peters-Stanley and Yin 2013), compared with only approximately one-third in 2006 (Hamilton *et al.* 2007: 47). Since 2007–2008, therefore, there has been a trend towards greater synergies between the voluntary and compliance markets – a convergence on a 'middle-ground' – with acknowledgement of the case against perfectionism in calculating greenhouse gas emission reductions.

In this chapter, I have suggested that in studying carbon offset markets it is crucial to pay close attention to the material world and role of technologies, which form a vital part of the "plethora of actors" (Sending and Neumann 2006: 668) in contemporary climate governance. Through drawing on theories of socio-technical transition, insights are provided into the dynamics of change, especially between different types of market. A key finding from the empirical case of carbon offsets is the existence of a form of market-based 'bottom-up' financial protection provided by the voluntary market innovation niche. Innovation leads to lower costs and less stringent regulation and therefore to an easier market within which to experiment. The voluntary offset market 'innovation niche' is, however, distinctive from existing examples in transitions scholarship of types of financial protection for innovation niches, which typically take the form of 'top-down' public sector grants and subsidies (Smith and Raven 2012). Further, the analysis in this chapter draws attention to the overlapping terminology in theories and concepts encompassed within the broad Part I theme of heterogeneous networks. In particular, the governmentality definition of technology refers to techniques, devices and practices, whereas the more STS-orientated definition encompasses engineered, material objects. The semantic confusion hints at a more substantial issue: the near-absence in governmentality of ideas about the conduct of technologies being managed; for governmentality is largely about the governance and self-control of people (Foucault 1991; Dean 1999). In contrast, STS theories do ascribe agency to technologies: these theories are centred on human-technology interactions at a micro-level, but miss the broader governance and policy focus.

The analysis here, above all, reveals the different ways that low carbon markets and economies can be made. Carbon offset markets represent one end of the spectrum of types of low carbon economy: they are a deliberately new type of market, purposefully designed with the primary objective of mitigating climate change. But despite their newness, they comprise a mix of older, pre-existing ideas and practices taken, for example, from the arenas of international development and global finance. 'New' carbon markets thus exhibit strong similarities to other, 'older' low carbon economies analysed in subsequent chapters. These older economies include those for housing and energy, where the issue of climate change has infiltrated and merged with long-established ways of doing, institutions and technologies, to which the analysis now turns in Chapters 3 and 4.

Notes

1 Parts of Chapter 2 have been published previously with Professor Diana Liverman, in: Lovell, H, and D. M. Liverman (2011). "Understanding carbon offset technologies". *New Political Economy* 15: 255–273. With thanks to Professor Liverman and Taylor and Francis Ltd. for kindly agreeing to me reusing aspects of this paper here.

2 A greenhouse gas is an atmospheric gas that absorbs and emits radiation, thereby influencing the temperature of the Earth. The 'basket' of six greenhouse gases defined within the UNFCCC Kyoto Protocol (1998) (see http://unfccc.int/resource/docs/convkp/kpeng.pdf) (Annex A) comprises: carbon dioxide (CO_2), methane (CH_4), nitrous oxide (N_2O), hydrofluorocarbons (HFCs), perfluorocarbons (PFCs) and sulphur hexafluoride (SF_6). For simplification, these greenhouse gases are often referred to simply as 'carbon'.

3 Our research at the UNFCCC meetings involved interviews (10) with offset organisations about what they were doing at the COP and why; participant observation at key business events hosted by offset organisations (and others) at the COP (e.g. Business Day in Bali); a survey of UN official side events (thirty in Bali; seventy in Poznan) to establish who was attending and how the discussion related to the main negotiations; plus informal discussions with offset companies and other relevant organisations in the corridors and at social events.

4 Additionality is defined in 3/CMP.1, Annex, paragraph 43 as: "A CDM project activity is additional if anthropogenic emissions of greenhouse gases by sources are reduced below those that would have occurred in the absence of the registered CDM project activity" (3/CMP.1, Annex, paragraph 43) (see http://cdmrulebook.org/84). Additionality is hence the requirement that the greenhouse gas emissions after implementation of a CDM project activity are lower than those that would have occurred in the most plausible alternative scenario (the 'baseline') if the CDM project had not proceeded.

5 This observation is made in relation to the period in question – 2007–2008 – but it is not the case from 2008 onwards, when the voluntary market implemented significant changes in systems of audit and regulation; see Hamilton *et al.* (2007); The Climate Group (2007).

Bibliography

Akrich, M. (1992). "The de-scription of technical objects". *Shaping Technology/Building Society: Studies in Sociotechnical Change*. W. E. Bijker and J. Law (eds). Cambridge, MA, MIT Press: 205–224.

Arthur, W. B. (1989). "Competing Technologies, Increasing Returns, and Lock-In by Historical Events". *The Economic Journal* **99**: 116–131.

Ascui, F. and H. Lovell (2011). "As frames collide: making sense of carbon accounting". *Accounting, Auditing and Accountability Journal* **24**(8): 978–999.

Baker & MacKenzie (2014). "CDM rulebook: what are small-scale projects?" Online at: www.cdmrulebook.org/152 (accessed 7 May 2014).

Boushel, C. (2014). "Assembling the taken-for-granted: carbon offsets and voluntary standards". Doctoral thesis, Edinburgh, University of Edinburgh.

Bumpus, A. (2009). "Understanding the materiality of carbon in carbon offset projects". Paper presented at the American Association of Geographers (AAG) Conference 2009, Las Vegas, 22–27 March.

Bumpus, A. and D. Liverman (2008). "Accumulation by decarbonisation and the governance of carbon offsets". *Economic Geography* **84**(2): 127–156.

Bumpus, A. G. (2011). "The matter of carbon: understanding the materiality of tCO_2e in carbon offsets". *Antipode* **43**(3): 612–638.

Callon, M. (1998). "An essay on framing and overflowing: economic externalities revisted by sociology". *The Laws of the Markets*. M. Callon (ed.). Oxford, Blackwell: 244–269.

Callon, M. (2009). "Civilizing markets: carbon trading between *in vitro* and *in vivo* experiments". *Accounting, Organizations and Society* **34**(3–4): 535–548.

Capoor, K. and P. Ambrossi (2007). *State and Trends of the Carbon Market 2007*. New York, NY, The World Bank.

Capoor, K. and P. Ambrossi (2009). *State and Trends of the Carbon Market 2009*. New York, NY, The World Bank.

Carbon Finance (2007a). "EcoSecurities' woes prompt CER rethink". Online at: www.carbon-financeonline.com/index.cfm?section=lead&action=view&id=10846 (accessed 20 November 2007).

Carbon Finance (2007b). "Let a thousand projects bloom". Online at: www.carbon-financeonline.com/index.cfm?section=features&action=view&id=10578 (accessed 17 May 2007).

Clean Development Mechanism (CDM) Policy Dialogue (2012). "Climate change, carbon markets and the CDM: a call to action. Final report of the high-level panel on the CDM policy dialogue". Bonn, UNFCCC.

Dean, M. (1999). *Governmentality: Power and Rule in Modern Society*. London, Sage Publications Ltd.

FERN (2005). "FERN briefing note June 2005 – carbon 'offset' – no magic solution to 'neutralise' fossil fuel emissions". Online at: www.fern.org/media/documents/document_884_885.pdf (accessed 12 October 2007).

Foucault, M. (1991). "Governmentality". *The Foucault Effect: Studies in Governmentality*. G. Burchell, C. Gordon and P. Miller (eds). London, Harvester Wheatsheaf: 87–104.

Friends of the Earth (FoE) (2009). "A dangerous obsession: the evidence against carbon trading and the real solutions to avoid a climate crunch". Online at: www.foe.co.uk/resource/reports/dangerous_obsession.pdf (accessed 10 December 2009).

Geels, F. W. (2004). *Technological Transitions and System Innovations: A Co-Evolutionary and Socio-Technical Analysis*. Cheltenham, Edward Elgar.

Hamilton, K., R. Bayon, G. Turner and D. Higgins (2007). "State of the voluntary carbon markets 2007: picking up steam". Washington, DC, New Carbon Finance and Ecosystem Marketplace.

House of Commons Environmental Audit Committee (2007). "The voluntary carbon offset market. Sixth report of session 2006–2007". London, The Stationery Office Ltd.

International Emissions Trading Association (IETA) (2008). "IETA's guidance note through the CDM approval process". Online at: www.ieta.org/ieta/www/pages/getfile.php?docID=2370 (accessed 18 December 2008).

Kollmuss, A., H. Zink and C. Polycarp (2008). "Making sense of the voluntary carbon market: a comparison of carbon offset standards". WWF-International, the Stockholm Environment Institute and Tricorona. Online at: http://assets.panda.org/downloads/vcm_report_final.pdf (accessed 2 June 2008).

Kossoy, A. and P. Guigon (2012). *State and Trends of the Carbon Market 2012*. Washington, DC, The World Bank.

Latour, B. (1992). "Where are the missing masses? The sociology of a few mundane artifacts". *Shaping Technology/Building Society: Studies in Sociotechnical Change*. W. E. Bijker and J. Law (eds). Cambridge, MA, MIT Press: 225–258.

Law, J. (1992). "Notes on the theory of the actor network: ordering, strategy and heterogeneity". Online at: http://comp.lancs.ac.uk/sociology/soc054jl.html (accessed 12 August 2014).

Li, T. M. (2007). *The Will to Improve: Governmentality, Development, and the Practice of Politics*. Durham, NC, and London, Duke University Press.

Lohmann, L. (2005). "Marketing and making carbon dumps: commodification, calculation and counterfactuals in climate change mitigation". *Science as Culture* **14**(3): 203–235.

Lohmann, L. (2009). "Toward a different debate in environmental accounting: the cases of carbon and cost-benefit". *Accounting, Organizations and Society* **34**(3): 499–534.

Lovell, H., H. Bulkeley and D. Liverman (2009). "Carbon offsetting: sustaining consumption?" *Environment and Planning A* **41**: 2357–2379.

Lovell, H. (2010). "Governing the carbon offset market". *Wiley Interdisciplinary Reviews: Climate Change* **1**(3): 353–362.

MacKenzie, D. (2009). "Making things the same: gases, emission rights and the politics of carbon markets". *Accounting, Organizations and Society* **34**(3–4): 440–455.

NC State (2014). "The past, present, and future of improved cookstove initiatives". Online at: https://sites.google.com/a/ncsu.edu/khopkarworldforestry/#TOC-Prior-Attempts-to-Disseminate-Cookstoves (accessed 14 June 2014).

Peters-Stanley, M. and D. Yin (2013). *Maneuvering the Mosaic: State of the Voluntary Carbon Markets 2013*. Washington, DC, and New York, NY, Forest Trends' Ecosystem Marketplace and Bloomberg New Energy Finance.

Plan Vivo (2007). "Plan vivo carbon management and rural livelihoods – projects". Online at: www.planvivo.org/fx.planvivo/scheme/projects.aspx (accessed 12 October 2007).

Power, M. (1999). *The Audit Society: Rituals of Verification*. Oxford, Oxford University Press.

Schot, J. W., R. Hoogma and B. Elzen (1994). "Strategies for shifting technological systems: the case of the automobile system". *Futures* **26**(10): 1060–1076.

Sending, O. J. and I. B. Neumann (2006). "Governance to governmentality: analysing NGOs, states and power". *International Studies Quarterly* **50**: 651–672.

SinksWatch (2007). "SinksWatch – an initiative to track and scrutinize carbon sink projects". Online at: www.sinkswatch.org/ (accessed 28 November 2007).

Smith, A. (2007). "Translating sustainabilities between green niches and socio-technical regimes". *Technology Analysis & Strategic Management* **19**(4): 427–450.

Smith, A. and R. Raven (2012). "What is protective space? Reconsidering niches in transitions to sustainability". *Research Policy* **41**(6): 1025–1036.

Taiyab, N. (2006). "Exploring the market for voluntary carbon offsets". London, International Institute for Environment and Development (IIED).

The Climate Group (2007). "VCS launch – a new quality assurance for the world's carbon market". Online at: http://theclimategroup.org/index.php/news_and_events/news_and_comment/vcs_launch_a_new_quality_assurance_for_the_worlds_carbon_market/ (accessed 19 November 2007).

Tyndall Centre (2008). "Programme one – informing international climate policy: how can international action on climate change be effectively developed after 2012?" Online at: www.tyndall.ac.uk/research/programme1/ (accessed 12 March 2008).

United Nations Reducing Emissions from Deforestation and Forest Degradation (UN REDD) Programme (2014). "About REDD+". Online at: www.un-redd.org/aboutredd/tabid/102614/default.aspx (accessed 7 May 2014).

United Nations Framework Convention on Climate Change (UNFCCC) (2012). "The CDM policy dialogue: background". Online at: http://cdm.unfccc.int/about/policy/index.html (accessed 19 March 2014).

United Nations Framework Convention on Climate Change (UNFCCC) (2014). "Press release – negotiations towards new universal climate agreement in 2015 get underway". Online at: https://unfccc.int/files/press/press_releases_advisories/application/pdf/20141403_adpclose.pdf (accessed 17 March 2014).

Unruh, G. C. (2002). "Escaping carbon lock-in". *Energy Policy* **30**(4): 317–326.

Wara, M. (2007). "Is the global carbon market working?" *Nature* **445**(8): 595–596.

3

IMPLEMENTING RESIDENTIAL DISTRICT HEATING

Key findings

- Climate change has emerged as a problem, and is implemented within, an already existing policy landscape for the UK social housing residential sector, including fuel poverty alleviation, social housing regeneration and welfare reform.
- Chapter 3 demonstrates the fragility of heterogeneous networks, as key actors (technologies and people) stabilising the market fail to function as planned or disappear, causing 'unravelling' and difficulty in meeting market objectives.
- Low carbon technologies are not reliable market actors – they are prone to destabilise or alter market operation – because they are typically new. This is evident with the new household metering technologies implemented at a housing development in Scotland, UK, which failed to provide accurate data on individual household heat consumption in a way anticipated by the Council, thereby putting pressure on the economic viability of the district heating.
- Chapter 3 reveals how establishing the economic credibility of residential district heating has been as important as establishing its sociotechnical credibility (i.e. how well it functions as a residential heating technology).

Introduction

The implementation of district heating into peoples' homes exemplifies the importance of emotions, feelings and other 'non-economic' factors in diverse low carbon economies. This chapter begins, therefore, by considering one of our interviewees, and the considerable impact that the new low carbon district heating system had for her. We spoke to Sandra on her doorstep, with her young son in her arms. For Sandra, the installation of district heating in her home – the housing development of Byford,[1] located near the centre of a major Scottish city

– has prompted a huge change in her lifestyle and day-to-day practices. No longer does she need to stay at her mother's house on the other side of the city during cold weather, nor hurriedly wash in the cold bathroom with its one-bar electric heater (one of only two sources of heat within the entire flat). District heating for Sandra, as for the other 1,500 tenants of Murray Housing Association living at Byford, has been truly innovative, transforming her capacity to heat her home at a relatively affordable rate and changing her everyday decision-making and activities at home and beyond. Much the same can be said for the tenants at Fordside – a smaller social housing development in another large Scottish city that is the main focus of this chapter, run by the local authority, where district heating was installed in early 2012.

In Chapter 3 the implementation of small residential district heating schemes in these two Scottish cities is used to demonstrate how low carbon economies are made and remade. As noted, the analysis concentrates primarily on the social housing development of Fordside, where district heating was installed in 2011. It also draws, as a comparison where appropriate, on findings from a second social housing development – Byford – where our interviewee Sandra lives, and where district heating was also installed during 2011.[2] The analysis shows how a low carbon economy has come into being in the context of Scottish housing – specifically, social (low income rental) housing. The implementation of new district heating systems has been financially enabled by a UK Government climate change and fuel poverty policy, enacted by energy utilities: the Community Energy Saving Programme (CESP). But it has also been driven by social housing modernisation reform – namely, the new 2015 Scottish Housing Quality Standards (SHQS) (Scottish Government 2013a, 2013b) – and its implementation has been influenced by the 2013 'bedroom tax',[3] a key UK Government welfare reform policy (see HM Government 2012). Tensions between these policy objectives are shown to have a number of practical ramifications for the installation and early operation of district heating, and these tensions have manifested most clearly with regard to debates about billing.

The detailed case study of Fordside tells us something about the particularities of implementing district heating, but also provides insights into the processes by which wider energy and heating economies operate and respond to change. The complex network of technologies (radiators, pipes, insulation, boilers, meters), institutions, financial models and householder energy practices integral to the making of a new low carbon district heating economy are identified, and the relations between them explored. What a low carbon economy means for Fordside is not yet settled; there are several unresolved issues that stem from different prioritisations of environmental and social policy objectives, but which manifest in economic debates. This chapter identifies how the material substance of district heating – the new networks of pipes, radiators and meters – plays an active role within the new economies of residential district heating. In other words, district heating is not just inert technology, but has acted in often surprising, unexpected ways to shape the decision-making about district heating and its implementation. These findings are broadly in keeping with a heterogeneous network approach to theorising the operation of economies and markets.

As noted, the chapter focuses primarily on the empirical findings from Fordside. Fordside is located in the old port area of a major Scottish city. Inefficient and expensive high carbon electric heating was replaced in the period 2011–2012 by more efficient, lower cost and lower carbon gas district heating, as part of a wider social housing regeneration initiative. Similar implementation of district heating took place in the inner city social housing development of Byford, also in the period 2011–2012. A heat interchange unit, new central heating radiators, controlled by a programmer and thermostat, and a meter have been installed in every home, in both cases replacing electric storage heaters. 'District heating' is, in reality, a catch-all term for a number of different technological configurations: a general point, which is observed in microcosm in Fordside and Byford. Fordside comprises a single multi-storey housing block with new heating from communal gas boilers, situated in the basement. In contrast, in Byford the Housing Association has invested in a Combined Heat and Power generator, located in a new energy centre, and an underground heat network has been installed to serve the entire housing development.

The chapter is structured as follows: first, the residential developments of Fordside and Byford are introduced, in order to demonstrate the different contexts into which new low carbon economies are implemented (for this, of course, has an effect on how things turn out – an obvious point, but one that is often overlooked in studies of innovation and change (Coenen and Truffer 2012)). It is noted that a key driver of the installation of district heating is a policy initiative from another (non-energy) policy domain: housing – namely, the introduction of the new Scottish Housing Quality Standards. Second, the case study of Fordside is considered in detail, in order to demonstrate the fragility of the heterogeneous networks that constitute this newly made (or remade) low carbon economy. A number of unexpected events occurred during the implementation and early operation of the district heating at Fordside, ranging from non-functional heat meters to the sudden absence of the Council Liaison Officer, which are explained and their implications examined.

Fordside and Byford: an introduction

Table 3.1 summarises several key characteristics of the two housing developments (including similarities and differences between them). Fordside and Byford comprise mostly social (low income rental) housing, with a relatively small number of owner occupiers (only three at Fordside; *c.*560 at Byford). Fordside comprises just over 200 homes, all flats. Byford is much bigger than Fordside at nearly 2,000 homes and contains a mix of different dwelling types (multi-storey and maisonettes).

One notable difference between the two housing developments is the type of landlord: in Byford the landlord is Murray Housing Association, a Registered Social Landlord (RSL), whilst at Fordside the landlord is the local authority (hereafter 'the Council'). Importantly for this analysis, the type of landlord has an influence on the governance of energy services. Fordside is unusual in the UK social housing sector, and, indeed, within the Council, in that the Council runs the energy services 'in-house', including individual consumption-based billing for its tenants. In other words, the Council has provided the new district heating and oversees the system and bills the householders directly for their usage (note that electricity is billed separately by utilities; the Council bills just for gas use – i.e. for space heating and hot water). In Byford, by contrast, the new energy services are provided by a private utility, who operates and maintains the generator and network and bills the tenants, under a long-term contract with the Registered Social Landlord – Murray Housing Association. This mix of public and private actors, operating at different scales, is typical of the liberalised and privatised energy sector in the UK (Graham and Marvin 2001).

Another notable difference between Fordside and Byford is their size and layout. Fordside is a much smaller social housing development, comprising a single block located within a constrained urban site. The site constraints led to a decision by the Council to install a suite of six large gas condensing boilers in a vacant ground floor location. In other words, the site of the existing building and its architecture determined to a large extent the type of district heating installed. At Byford, in contrast, there was the space within the perimeter of the housing development to build a new dedicated energy centre building, which houses a Combined Heat and Power (CHP) unit. This is a gas-fired power plant that produces electricity as well as heat. The heat is supplied to Byford households (which required a network of new heat distribution pipes to be laid all around the site, see Plate 3.1), and the electricity is sold into the power networks, forming a small part of the utility's total electricity generation.

A final difference worth highlighting from Table 3.1 is the existence of a concierge service at Fordside; there is no equivalent at Byford. As discussed later in the chapter, the twenty-four hour Fordside concierge – a team of building managers based in an office at the entrance to Fordside, in operation since 1993 – has played a role in stabilising the new hybrid network of district heating.

Table 3.1 Key general characteristics of Byford and Fordside

	Byford	*Fordside*
Location	*Major Scottish city (#1)*	*Major Scottish city (#2)*
No. of properties	*c.* 2000	*c.* 200
Type of landlord (city-wide)	In this Scottish city, the City Council is no longer a social housing provider: housing stock was transferred to individual housing associations (social landlords) in the early 2000s, including smaller social landlords such as Murray Housing Association (HA), which owns and runs Byford.	In this Scottish city, voting was against housing stock transfer; the Council retains responsibility for most multi-storey social housing in the city.
Built environment	The Byford estate was built in the 1960s by the Scottish Special Housing Association (SSHA) as an 'elite' scheme, with around 1,900 flats and maisonettes.	Fordside is a single nine-storey block, with *c.* 200 flats, built by the Council in the 1960s. The flats are mainly deck access, from a corridor, except on the ground floor, where each has a separate front door.
Type of tenure	Approximately 560 houses (mostly maisonettes) were bought by their occupiers under the UK Government 'right to buy' scheme.	Fordside has remained in continuous Council ownership since it was built. A very small number of the flats (#3) have been bought under the UK Government 'right to buy' scheme.
Poverty and social problems	Poverty is more entrenched at Byford than in earlier periods; drug and alcohol problems are perceived to have increased.	Twenty-four hour concierge office introduced in early 1990s in response to social problems; drug and alcohol problems are perceived to have lessened since then.
Landlord-tenant relationship	Relative dissatisfaction and lack of trust of Murray HA among tenants. Murray HA is perceived as not sufficiently interested in tenants' welfare and concerns.	Reasonable satisfaction with the Council, but some difficult landlord-tenant relationships.
Financing	Murray HA struggled with limited finance – they bought the estate as part of a stock transfer, so have had no finance for regeneration since the estate was acquired in 1994. Funds for some improvements became available only recently, when tenants agreed (by vote) to a 9.5% rent increase. Hence, the initial phase of ownership resulted in cost-cutting and less contact with tenants than under the management of the previous housing association. Since the district heating installation, eight housing officers, each with responsibility for a specific area of the estate, have been introduced to work with tenants on issues such as managing finance, rent, dealing with heating costs.	The Council housing services are less financially constrained; they are able to run concierge services and had an officer working on local liaison during the refurbishment. The Council has increased rents routinely, with many tenants on low incomes or unemployed receiving state housing benefits. A rolling programme of housing stock upgrade has therefore proceeded.

Plate 3.1 Installation of district heating pipework at Byford, early 2012 (source: Professor David McCrone, University of Edinburgh).

The decision to install district heating

In both Fordside and Byford the impetus to implement district heating stemmed in large part from changes required under Scottish housing policy – namely, the introduction of new Scottish Housing Quality Standards (SHQS). The main catalyst was not, therefore, directly to do with climate change. The SHQS were upgraded in early 2011 and compliance with the new standards (across all areas, including energy efficiency) is required by social housing landlords by April 2015. All Scottish social housing landlords are therefore in the process of renovating their existing homes to meet the SHQS. Annex C (Energy Efficiency) of the SHQS requires 'full' and 'efficient' central heating (Scottish Government 2013a): the outdated and inadequate electric storage heating present in Fordside and Byford was not compliant. Further, because of the solid concrete housing construction method used in both developments, there was some anxiety about installing a new heating system that required piping gas through the buildings. In the UK there is concern that piped gas is not safe within the type of multi-storey concrete building found at Fordside and Byford, because of an explosion and partial collapse of a concrete tower block – Ronan Point in London in 1968 – which was caused by a boiler gas leak from a flat located in an upper storey. District heating therefore provided a way of avoiding piped gas, because instead it involves hot water that is circulated around the buildings.

A lead person involved in the decision to adopt district heating at Byford explains the decision-making process for them as follows:

> . . . the cost and the issues associated with installing nearly 1,500 combi-boilers in Byford, just weren't practical, it wasn't possible to do that. Plus the tower blocks, strictly speaking one can pipe gas into tower blocks, but post-Ronan Point nobody really wants to do that.
>
> (Interview, project manager, Byford, October 2013)

This illustrates how the accident at Ronan Point still resonates and has an influence in new energy technology decision-making for this type of 1960s solid concrete multi-storey housing, despite the passage of time that has elapsed.

The new SHQS also stipulates standards for bathrooms and kitchens (Scottish Government 2013b), and at Fordside a decision was taken to upgrade these facilities at the same time as installing district heating. At Fordside district heating is consequently viewed by householders, and by many of those at the Council, as bundled up with the housing refurbishment. Further, a dedicated member of staff at the Council was responsible for overseeing the process for the Fordside residents (about which, more later in this chapter). This broader approach to district heating installation at Fordside – coupling it with other housing renovations – stems in part from a relatively more generous institutional funding environment within the Council, as compared with Murray Housing Association who runs Byford (see Table 3.1), as well as the management of energy 'in house' with the Council for this particular housing development, allowing for greater co-ordination of energy innovation with housing renovations.

The making of new low carbon economies in the two case studies goes hand-in-hand, therefore, with social housing modernisation. However, there is a second policy objective which district heating is also closely bundled up with – namely, alleviating fuel poverty. Fuel poverty has been an extensive problem in both Fordside and Byford and was a related driver for the heating system upgrade, because electric heating is expensive to run. Scotland has an ambitious target of eradicating fuel poverty by 2016, set by The Housing (Scotland) Act 2001 as a statutory duty on the Scottish Government (Scottish Government 2001). The Community Energy Saving Programme (CESP) – in operation between 2009 and 2012 across the UK – which provided core funding to Fordside and Byford, has explicit dual policy objectives of mitigating fuel poverty as well as climate change. Negotiated as part of the utilities' agreement with the UK Government, CESP finance had to be spent in designated low income areas, for which Fordside and Byford both met the criteria. CESP used finance from the energy utilities, which was, in turn, added as a charge to all customer energy tariffs. Fordside was awarded £750,000 in CESP funding, and Byford £15 million (the largest CESP award in the UK; personal communication (27 July 2012)). In total, the CESP provided funding for approximately 23,000 new district heating/Combined Heat and Power 'measures' in households across the UK (with, on average, just under two 'measures' per property; see OFGEM 2013). Under the CESP rules, energy efficiency building retrofits (roof insulation, cladding, etc.) had to be undertaken at the same time as the new district heating was installed; and both housing developments received funding for this energy efficiency work as well under the CESP.

Even this brief outline of the context for district heating in these two housing developments highlights the multiple objectives at work: a climate change policy objective (evident most clearly in CESP), housing stock modernisation (as evidenced by the new SHQS), as well as an ambitious Scottish Government target to eradicate fuel poverty by 2016. So, climate change has emerged as a problem, and is implemented within, an already existing, crowded policy landscape. Further, district heating is implemented in places which have an existing socio-technical infrastructure. The type of building construction and architecture, size and layout of the site, its social history, tenant composition, prior methods of heating the home and so on have all influenced whether, and how, new district heating low carbon economies have been made.

Fordside: the work of holding together a fragile heterogeneous network

Having explained the context of Fordside and the decision to implement district heating, using the social housing development of Byford as a comparison, the analysis now turns to consider in more detail the relations between people and institutions at Fordside (the Council, householders, installers, the concierge and the Council Liaison Officer) and the material elements of the district heating (the new radiators, heat meters, wiring and heating control panels, etc.) using a heterogeneous networks approach. The Fordside case study illustrates the work required to build a new heterogeneous network and keep it together, for there were several unexpected issues that arose as one or more vital components failed to function as planned. The case of Fordside therefore illustrates the fragility of economies in the process of being reformed because of climate change, uncovering the constant work required 'behind the scenes' to maintain stability. Further, the care with which the new district heating heterogeneous network was assembled at Fordside shows the degree of forethought and crafting that went into conceiving of this particular market and then putting it into practice. The crafting of the market was fundamental to its early success – namely, the smooth implementation process and high levels of tenant satisfaction we observed in our interviews in early 2012, just weeks after the new heating 'went live'. The development of new district heating at Fordside and its implementation was carried out with close attentiveness to the housing development's characteristics: its institutions and history, tenant mix, type of building, landlord–tenant relationship and so on.

A large part of the rationale behind the Council's close attention to the early stages of district heating planning and implementation stemmed from the need to gain explicit, signed agreement from tenants to install district heating in their homes. Under the 2001 Housing (Scotland) Act (Part III, section 236), the landlord (in this case, the Council) does not have powers to force existing tenants to agree to new types of heating such as district heating: it is a matter of choice for the tenant (Scottish Government 2001). The landlord is allowed access into homes to install the basic pipework, but existing tenants – under the Tenancy Agreements they signed – are not obliged to then accept a change in their heating. This created a potential problem for the Council: if a large number of Fordside tenants had refused to agree to the new district heating, then this would have put the project at risk of not going ahead. This is because the district heating boilers (located in the Fordside basement) had to be sized based on the maximum load – i.e. all of the *c.*200 households at Fordside. In other words, the (large) upfront capital cost of installing the district heating was going to be the same, regardless of how many tenants signed up. So, it is not hard to see that under a scenario of, say, a 50 per cent sign-up rate, the new district heating would be unlikely to create sufficient revenue for the Council. Therefore, one of the first and most important tasks for the Council in creating a new district heating heterogeneous network at Fordside was successfully enrolling its tenants. The approach the Council took was one of openness and information provision.

Crucially, a decision was made early on in the process by the Council to appoint a full-time staff member, first employed in 2010, to act as a liaison between the residents, contractors and the Council. The Council Liaison Officer, who was seconded from the local Tenants Federation, carried out a number of tasks that greatly facilitated the uptake of district heating and the process of installation on site. For example, she wrote and distributed regular monthly newsletters sent out to all Fordside residents about the refurbishment and district heating (starting in December 2010 and ending in December 2011). She also developed a comprehensive information pack at the start of the resident consultation process (October 2010) and organised a dedicated residents meeting (19 October 2010). Further, she arranged for a demonstration home to be set up at Fordside for residents to visit and choose colour schemes for their refurbishment and to see the new radiators, heat exchange unit and heat control panels.

Fordside district heating and installation work started in January 2011 and took fourteen months to fully complete: the timetable was not overly rushed, and the various stages were planned out in some detail. Particular care was taken by the Council Liaison Officer throughout in communicating with residents and notifying them of what was going to happen, as an officer at the Council explained: “She was visiting tenants explaining how the system would work. . . . She was meeting them regularly, writing to them regularly, answering their questions over the phone regularly. She was the link between the Tenants Federation and the Council” (Interview, Council officer, July 2013). The Council Liaison Officer was recruited in large part because of her ten-year experience as a development worker at the local Tenants Federation, and, as the interview extract indicates, she drew on this experience to expertly manage relationships with Fordside tenants, with frequent phone calls and visits to inform and guide tenants unsure about whether or not to accept the new heating. Her work was complemented by the concierge service that operates at Fordside. The concierge comprises a twenty-four hour team of trained building managers, who help manage the block in various ways and who have provided ad hoc advice and information to tenants about the new district heating, acting as a type of intermediary.

A key factor for the residents of Fordside in deciding whether or not to accept the new heating system was, of course, the poor quality and expensive system of heating that was in place previously; in most cases, just two poorly functioning electric storage heaters per property. There was hence a much lower risk for the householders in opting to try something new. This is a good example of why attention to place and local sociotechnical context is important in further developing our understanding of precisely how new low carbon economies are made: a different sociotechnical starting point is likely to have considerably altered how, and indeed whether, district heating was implemented. Further, Fordside only has a very small number of owner occupiers (3), and there is some evidence to suggest that Fordside was chosen by the Council as the first multi-storey to install district heating in the city because of its low proportion of private owners (various interviews, 2012). Private owners were judged to be more likely to refuse district heating, because, unlike Council tenants, they were required to pay the upfront cost of installation (set by the Council at £2,000). One out of the three private owners at Fordside did, in the end, agree to have the district heating installed.

Understandably, a key issue raised by Fordside residents in deciding whether or not to sign up for the district heating was the expected cost of the heat. The majority of discussions the Council Liaison Officer had with tenants were about cost (personal communication, meeting with the Council, 17 January 2012). Again, the prior heating infrastructure at Fordside in effect facilitated the transition to district heating, because not only was it poor quality, but it was also expensive. The Council further influenced householders’ decision to adopt district heating by giving tenants the promise of low fixed price heat for a year. Tenants were told they would eventually pay for what they use, but for the first year of operation of the district heating (April 2012 to March 2013) there was to be a flat rate charge of £1 per day for space (central) heating and hot water, resulting in savings in bills for households. Thereafter, from April 2013 onwards, the Council pledged to provide the heat to residents at Fordside 'at cost' – i.e. the Council itself was not seeking to make a profit from running the district heating at Fordside, but merely needed to break even financially. Additionally, for the first few months of the district heating operation (January–April 2012), whilst the system was still being commissioned and tested, it was agreed that no charges at all would be made to residents for the heat that they used.

So, through the work of the Council (especially the dedicated Council Liaison Officer), a robust network of relations was fostered during the early planning and implementation of district heating at Fordside. Crucially, this network of actors and relations was not totally new, but rather built upon and incorporated a number of diverse, already existing things, people and institutions, including: the concierge, the extensive problem of fuel poverty at Fordside, the low number of owner occupiers (3), the building type (1960s solid concrete), the 1968 Ronan Point gas explosion, the substandard electric heaters that were in place and so on. Conceptually, this attention to what was there before the district heating matters, for it substantiates the point that low carbon economies are not built from scratch. The heterogeneous network of district heating at Fordside was carefully and skilfully crafted, drawing together this mix of people and things, both new and old. The Council was successful in signing up the vast majority – 92 per cent – of the approximately 200 households at Fordside.

However, a number of unexpected things happened subsequently that led to a partial unravelling of this carefully constructed heterogeneous network – namely, the sickness and death of the Council Liaison Officer; faults with household metering; the UK Government introduction of a social housing 'bedroom tax'; and issues with the heating control panels. These distinct but interlinked events, policies and technical problems manifested most tangibly in disputes over billing, especially during the period in early 2013 towards the end of the '£1 a day' fixed price billing for heat. Each of the 'unravellings' are briefly described, and their impact analysed.

The Council Liaison Officer

Arguably one of the most important nodes in the heterogeneous network was the Council Liaison Officer. As explained above, she was employed to work by the Council in 2010, on secondment from the local Tenants Federation, with the purpose of facilitating tenant sign up to the new district heating at Fordside and managing the installation process. With her background on tenant committees, she had the experience and skills to do this job to a very high standard. As outlined, she undertook a range of activities to provide information to the Fordside tenants about the district heating and bathroom and kitchen upgrades. However, in early 2012 she was taken ill and sadly died in mid-2012. Her unexpected death, quite aside from other consequences, led to parts of the network unravelling once she was no longer looking after those relations. In particular, two technologies in the network failed to act as planned – the household heat meters and heating user controls – and in her absence these problems were not noticed and dealt with in a timely way. These 'misbehaving' or unruly technologies contributed to the most significant problem to emerge as a consequence of the Liaison Officer no longer being part of the hybrid district heating market, which was the belated identification of financial problems in the Fordside heterogeneous network (about which, more later in this chapter).

Damage to the household heat meter wiring

When the new heating was installed at Fordside, each household was fitted with an individual heat meter, designed to take half hourly meter readings from the home. These meters were installed in order to provide the Council with initial information (during the first year of district heating operation, 2012–2013) about each households' consumption of heat and were subsequently to be used for calculating individual household bills, based on actual consumption. As noted, for the first year of operation (April 2012 to April 2013) it was agreed that all tenants would be charged a flat rate of £1 a day, in order to facilitate sign up and reassure households about their bills. The £1 a day charge to households was based on Council estimations that this would cover their cost of providing heat (Interview, Council officer, July 2013). Therefore, the metering data had no direct financial consequences for the tenants or Council during 2012–2013, but it was intended, rather, as a way to check the operation of the district heating and as a means of providing the Council with an early warning of significant over- or under-consumption of heat by tenants.

However, it was not until late 2012 that it was discovered that the heat consumption data was not, in fact, being collected from the meters as originally planned: data was only available from approximately half the households (Interview, Council officer, July 2013). Reasons given for the heat meters not providing data included communication errors between the meters and the central database, and vacant flats (Various interviews with Council officers, 2012, 2013). In addition, one apparent source of the problem for a small number of households (9) was identified by a Council officer to be damage to the wiring leading to and from the individual household meters.[4] With the Council Liaison Officer absent, these problems with the heat meters went unnoticed during the start of the first crucial heating season (October to March in the UK). Meanwhile, vital data on heat consumption for half the flats had been lost, with no possibility of recovery. It was only once the meters were up and running again for the majority of flats by

December 2012 that a situation of significant and persistent over-consumption of heat by a small number of tenants was identified by the Council. Coupled with higher than anticipated distribution losses from the network of heat pipes circulating the building, the result was that the Fordside district heating unexpectedly ran at a significant loss in its first year of operation.[5] In other words, it was discovered that the £1 a day charge to householders for their heat during 2012–2013 had been insufficient to cover operating costs: the Council were far from breaking even financially, which had been their original aim. Because of the absence of metering data, this financial loss in the Fordside district heating market was not identified until around the time when households were due to move to consumption-based billing (April 2013).

Once the Council had the consumption data, they promptly instigated a series of home visits with the small number of tenants identified as persistent over-users of heat. These home visits, which took place in May 2013, brought to light another key player in this heterogeneous network: the household heat controls. The controls were installed in every home at the time of implementing the district heating, but, like the gathering of data from the new heat meters, failed to operate quite as planned.

Household heat controls: a complex control panel with poor instructions

How the heat control panels were used (or misused) by Fordside householders demonstrates again the capacity of materials within markets to 'act back' – i.e. to behave in unexpected ways. Whilst a seemingly innocuous technology, the heating control panel placed in each home when the new district heating was installed was startlingly new to many householders (see Plate 3.2). A centralised, programmable control panel for heating and hot water was not something they had previously encountered with their electric storage heaters. The electric storage heaters came on automatically each night, making use of off-peak electricity, and then released their stored heat gradually throughout the day. Each electric storage heater had its own controls, and there was no centralised control panel or thermostat to regulate space heating. Our research with householders at Fordside, and the findings from the Council's one-to-one visits with persistent over-users, indicated that part of the reason for higher than expected consumption stemmed from this transition to new, unfamiliar heating controls.

In interviews, residents described to us their difficulties in understanding the controls, with one commenting that the instruction manual is "...like reading Chinese..." (Interview, April 2013), and many asking friends and family or the concierge for help in setting the heating and hot water to come on and off on an as needs basis: "If I get cold, my daughter showed me how to boost it" (Interview,

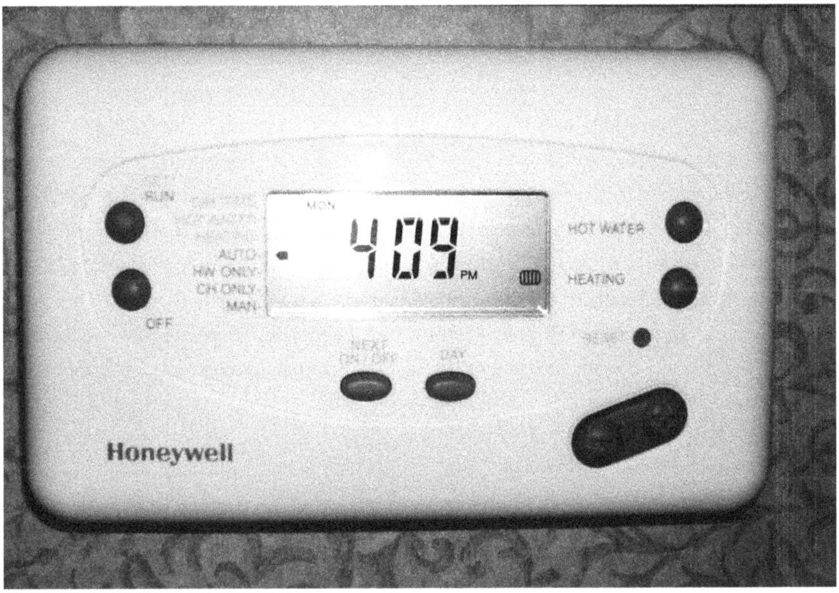

Plate 3.2 The heating control panel installed at Fordside (source: the author).

April 2013). We observed a multitude of different ways of using the control panel, with some residents programming it so the central heating and hot water came on and turned off at regular times, while others just turned it on manually each time they needed heat. For example, one householder explained to us how he manually sets the heating to 'boost' to control the temperature in the flat (in the process setting the thermostat unnecessarily high; an inefficient way to manage the heating):

> When it goes to 30°C . . . the wee fire will appear on the screen, that's fire on the radiator but if you don't see any fire coming on the screen sometimes I have still got to press it, press it, press it until that thing [fire] appears, so as soon as it appears I know that [the heat will come on].
>
> (Interview, April 2013)

One of the Council officers involved in the 'over-user' home visits explained the situation as follows:

> . . . the problem is they didn't teach any of these tenants how to use the programmer system. The technology only works in the right hands. You could put me in the cockpit of a jetfighter but I wouldn't be any good at it. Good technology only works in the right hands.
>
> (Interview, local authority officer, July 2013)

The Council Liaison Officer had planned to provide all householders with a DVD explaining how to use the programmer, as well as easy-to-read instructions, but neither materialised in her absence. New relations in the heterogeneous network between the householder and their district heating system were hence never properly forged and embedded. In a further twist to the story, one unforeseen element was what happened to the heating controls if there was a power cut – i.e. the electricity was disconnected in a home. The particular model of control panel installed at Fordside reset to a factory default of 'maximum heat' when electricity supply was restored. Whilst this might at first glance seem like a minor issue – power cuts are a fairly rare event in most UK households – in the particular setting of Fordside it became highly significant. This is because a number of tenants struggle to pay their bills on time (including utility bills), so electricity disconnections are therefore actually quite common. According to the Council, the resetting of the heat controls to the factory default of 'maximum' after an electricity disconnection (with the controls then left unaltered by the householder) is one of the reasons behind the over-consumption in a small number of homes (Interview, Council officer, July 2013).

The UK Government 'bedroom tax'

A final unexpected but nonetheless influential event was the introduction from April 2013 of a new UK Government welfare policy that effectively charges social housing tenants extra rent for having unoccupied bedrooms in their home, by reducing housing benefits (HM Government 2012). It has been a controversial policy, with much protest, and many tenants (UK-wide) have refused to pay the extra charge, including at Fordside. The relevance for district heating is that this so-called 'bedroom tax' was implemented at the same time as the Council was preparing to bill householders at Fordside for their actual consumption (i.e. the transition period away from the '£1 a day' fixed charge). Further, the method of payment for heat at Fordside by the householder is via a 'top-up' on their rental bills. With many householders at Fordside already refusing to pay another new additional charge on their rent (if they have a spare bedroom), it seemingly encouraged them to consider not paying any extra for their heat as well, as this interview extract illustrates:

> Now the [residents] meetings are about the Council Tax, for the bedrooms. We have got to pay £52 a month on top of that [the heat bill]. I am not paying that, I am not paying that . . . I'd say 50% of the building are single people, or married couples, [so] they only need the one bedroom . . .
>
> INTERVIEWER: So are you expected to pay that with the heating then?
>
> Yes, we have to pay £30 a month for the heating, and extra for our

> electricity on top of that, *and* £52 a month for the extra bedroom. It's not happening.
>
> (Interview, April 2013)

So, from the householder's perspective, the two simultaneous changes in their billing – from fixed rate to consumption-based bills for heat, as well as the 'bedroom tax' – were linked together, and for some residents they were prepared to protest by withholding payment for both.

The outcome: problems with billing

A key way these unexpected 'unravellings' of the hybrid district heating market at Fordside manifested has been through issues with the billing of householders for their heat. At the time of writing (2014), billing for the district heating has still not been resolved. It appears that householders have mostly had adjusted heat bills for the period since April 2013 (with some receiving rebates, and others bills for further payment), but there is as yet no regular system of consumption-based billing in place. Further, the rebates and bills that householders did receive during 2013 lacked any breakdown of consumption showing how their bill was calculated – i.e. individual household consumption data was not being provided (Householder interview, March 2014). The Council is in the process of developing a new Homes and Energy Strategy, with an associated Investment Plan, and its experience of implementing district heating at Fordside has informed their decision-making. Crucial to the Council's decision about which heating technology to invest in in the future is the financial viability of district heating, as an interviewee explained to us: "there's a bit of a financial hole at Fordside and that needs to be ironed out before we'll get the go ahead to move on to other projects and apply for other grants to install these things" (Interview, Council officer, July 2013). Hence, there is considerable pressure on the team within the Council with responsibility for Fordside to make district heating work, and, crucially, the criteria for 'what works' is that it is cost-effective. In other words, in making this important decision about new forms of heating provision in social housing across the city, the economic credibility of residential district heating appears to be as important for the Council as its sociotechnical credibility (i.e. how well it functions as a residential heating technology).

Summary and conclusions

This chapter explores the fragility of heterogeneous networks, using the example of residential district heating. The analysis in this chapter demonstrates how low carbon markets emerge in a complicated way – they are not neatly made – but instead draw in and reframe existing ways of thinking and doing. Further, Fordside provides a nice example of what happens when key actors (technologies and people) that are stabilising a new low carbon heterogeneous network fail to function as planned or disappear, causing 'unravelling' and difficulty in meeting planned objectives. It is observed that low carbon technologies are not especially reliable market actors – they are prone to destabilise or alter market operation – because they are typically new and therefore relatively untested. We see this played out, for instance, with the new household metering technologies at Fordside, which failed to provide accurate data on individual household heat consumption in a way anticipated by the Council, thereby putting pressure on the economic viability of district heating at Fordside. The example of district heating implementation in Scottish social housing also demonstrates how climate change has emerged as a problem, and is implemented within, an already existing policy landscape for the UK social housing residential sector, including fuel poverty alleviation, welfare reform (the 'bedroom tax') and housing regeneration and modernisation.

Notes

1 Note that the two housing developments under discussion in Chapter 3 have been anonymised. Byford and Fordside are pseudonyms and are not real places. Any similarity to existing places or housing developments called 'Byford' or 'Fordside' in Scotland or elsewhere is unintentional.

2 The chapter is based on empirical research conducted at two social housing regeneration projects in Scotland in the period 2011–2013. The research data comprises: semi-structured interviews with project developers and other organisations involved in establishing and running the district heating (10); longitudinal interviews (2011, 2012, 2013) with a small sample of householders at Fordside (12); and a structured face-to-face repeat survey (2012, 2013) of 200 (10 per cent) households at Byford. This research forms part of a four-year Research Councils' UK (RC-UK) funded project, "Heat and the city". "Heat and the city" comprises a multi-disciplinary research team examining the complex questions confronting society in relation to the future provision of low carbon heating in urban areas. The fieldwork undertaken for the residential component of "Heat and the city" has encompassed the users of district heating (the householders), as well as the 'upstream' organisations involved in arranging and delivering the new district heating within our two case studies.

3 The so-called 'bedroom tax' is a new UK welfare reform policy introduced on 1 April 2013 that means that all UK social housing tenants who have a spare bedroom in their property have had their level of government housing benefit reduced; see www.housing.org.uk/policy/welfare-reform/bedroom-tax/ (accessed 13 June 2014).

4 There is confirmed evidence that the wiring in three pram stores was found to be damaged, but it is not certain what the cause of this damage was – in particular, whether it was deliberate or accidental. It may have been caused by accident as tenants moved their things back into the storage areas after the building work was complete. The meter wiring in each pram store serves three flats, so a total of nine flats had their heat meters affected in this way.

5 Estimated at circa £30,000, based on estimate of £100,000 cost of gas supply per year (Interview, Council officer, July 2013), against 194 tenants with district heating paying £360 per year in bills.

Bibliography

Coenen, L. and B. Truffer (2012). "Places and spaces of sustainability transitions: geographical contributions to an emerging research and policy field". *European Planning Studies* **20**(3): 367–374.

Graham, S. and S. Marvin (2001). *Splintering Urbanism: Networked Infrastructures, Technological Mobilities and the Urban Condition*. London, Routledge.

Her Majesty's (HM) Government (2012). "Welfare Reform Act 2012". Online at: www.legislation.gov.uk/ukpga/2012/5/contents/enacted/data.htm (accessed 10 February 2014).

Office of Gas and Electricity Markets (OFGEM) (2013). "The final report of the Community Energy Saving Programme (CESP) 2009–2012". Online at: www.ofgem.gov.uk/ofgem-publications/58763/cesp-final-report-2013final-300413.pdf (accessed 12 August 2014).

Scottish Government (2001). "The Housing (Scotland) Act (2001)". Online at: www.legislation.gov.uk/asp/2001/10/pdfs/asp_20010010_en.pdf (accessed 11 March 2014).

Scottish Government (2013a). "Scottish Housing Quality Standards – annex C 'must be energy efficient'". Online at: www.scotland.gov.uk/Topics/Built-Environment/Housing/16342/shqs/AppendixC (accessed 20 March 2014).

Scottish Government (2013b). "Scottish Housing Quality Standards – annex D 'must have modern facilities and services'". Online at: www.scotland.gov.uk/Topics/Built-Environment/Housing/ 16342/shqs/AppendixD (accessed 20 March 2014).

Part II

FRAMING

In Part II of *The Making of Low Carbon Economies*, I turn to consider the concept of framing and explore how framing has been an integral process in making new, and reforming existing, low carbon markets and economies. Frames are defined as a lens on the way that problems are viewed, discussed and resolved:

> . . . a way of selecting, organising, interpreting and making sense of a complex reality to provide guideposts for knowing, analysing, persuading and acting. A frame is a perspective from which an amorphous, ill-defined, problematic situation can be made sense of and acted on.
>
> (Rein and Schon 1993: 146)

A key function of frames is to claim ownership, and frames are, hence, inherently political (Rein and Schon 1993; Hajer 1995; Hastings 1999; Jacobs *et al.* 2003; Bickerstaff *et al.* 2007).

Key insights provided by framing for the study of economies and markets include:

- Framing a new issue makes action possible – it sets the boundaries for the problem and therefore its solutions.
- Framing involves discourse, but also practices and materials.
- Framing is highly political, as the actor who defines a problem or situation gains authority over it.
- Dominant frames are mostly taken for granted and not explicitly recognised.
- Recognising the existence of frames encourages learning.
- Framing is an active process, it is never complete.
- Attention to the fragility of frames highlights the possibility of moments of significant change.

The concept of framing is interdisciplinary: it has its origins within Foucauldian scholarship, as well as public policy analysis, and has also been applied and further developed within economic sociology, environmental studies, human geography and STS (see, for example, Rein and Schon 1993; Abolafia 1998; Hoffman and Ventresca 1999; Laws and Rein 2003; Lovell 2004; Lohmann 2005, 2009; Callon 2007). In this introduction to Part II, I explore a number of insights that framing provides for the study of (low carbon) economies and markets, integrating ideas from across these different disciplines.

The first insight is that framing is an essential precursor to action. In the field of public policy analysis, framing is about setting boundaries around a policy problem, and thereby making the conception and implementation of policy solutions – i.e. action to resolve it – possible (Sharp and Richardson 2001). Similarly, for Callon and others working within economic sociology, frames allow market agents to calculate and make sense of a market by drawing boundaries around it; classifying and policing what is in and out of a market (Callon 1998; Barry and Slater 2007). Framing is thus seen as particularly relevant in situations where new policy problems (such as climate change) arise and where there are high levels of uncertainty regarding how to respond. Framing defines the problem (and therefore also its solutions) by structuring the terms of the debate, foregrounding certain forms of knowledge, expertise and practice as relevant and setting limits on what action is judged to be appropriate.

Framing is hence a valuable theme through which to consider the making of low carbon economies, because of its particular relevance to analysis of new areas of activity. It is for this reason that framing has been used within a number of studies of the emergence of, and response to, new environmental problems (Dryzek 1997; Sharp and Richardson 2001; Hajer and Wagenaar 2003). A consideration of framing sheds light on how something new like a low carbon market is forged out of, and made distinct from, existing practices and modes of operation. As Hajer and Versteeg (2005: 182) observe: "Environmental debates often take place in a situation of institutional ambiguity, in which there are no generally accepted rules and norms according to which politics is to be conducted and policy measures are to be agreed upon". And, despite nearly two decades of sustained effort to mitigate climate change, there still remains considerable "institutional ambiguity" regarding it. The chapters within Part II provide contrasting examples of the process of framing, illustrating the role of framing in making sense of climate change in specific (and quite different) policy and technology contexts: Chapter 4 is about the low carbon reframing of existing housing economies, whereas Chapter 5 is about a new area of professional activity brought into being by climate change: carbon accounting.

Second, frames comprise discourse ('rationalities'), as well as practices and technologies. The theme of framing is central to operationalising aspects of Foucault's ideas, particularly that of governmentality, which concentrates on how objects of government are defined and how problems (and their solutions) are framed (termed 'rationalities'), as well as how they are governed through 'technologies' (Dean

1999). The idea that frames are constructed and sustained through discourse as well as practices and technologies is also evident in economic sociology and public policy analysis of framing (Rein and Schon 1993; Callon 1998; Bickerstaff *et al.* 2007). The concept of framing hence has synergies with the sociotechnical concept of heterogeneous networks considered in Part I (see the introduction to Part I). Such an approach to framing is especially relevant in thinking about low carbon economies, because it brings to the fore the possibility that low carbon technologies can themselves influence how climate change is conceived of, discussed and operationalised as a problem: there is a two-way relationship between technologies and discourse. As leading scholars have observed (Lohmann 2009; Boyd *et al.* 2011; Lövbrand and Stripple 2011), the 'measure to manage' rationality that pervades carbon markets and low carbon economies can itself limit climate mitigation action to a relatively narrow band of technologies and activities that are cost-effective and can be measured and monitored. An economic framing of climate change mitigation thus places boundaries on what is viewed as credible, acceptable and workable. But, crucially, for Foucault and others, these boundaries are always subject to change. They can be negotiated and remade ('reframed') in quite different ways. This is demonstrated in Chapter 4, for instance, when examining how pioneering sustainable homes built in the UK encouraged the UK Government to further develop its ideas about low carbon housing, leading to the introduction of a radical, new zero carbon homes policy.

Third, dominant frames are so pervasive they are often not noticed. It is an observed feature of frames that they are seldom acknowledged and are often invisible to those using them, as Rein and Schon explain:

> Although frames exert a powerful influence on what we see and neglect, and how we interpret what we see, they are, paradoxically, difficult to assess. Because they are part of the natural, taken-for-granted world, we are often unaware of their role in organising our preconceptions, thoughts and actions.
>
> (1993: 151)

By explicitly acknowledging and highlighting frames, it is argued that there are significant opportunities to encourage constructive learning and policy change (what Rein and Schon (1993) term "frame-reflective discourse" (150); see Etzion and Ferraro (2010) for an example). We can apply this insight to the economic (neoliberal) framing of many environmental policy issues, including climate change. Climate change has primarily been framed as a problem that has been caused, and needs to be solved, through markets and economies. For example, Lord Stern, commissioned by the UK Government to examine and make recommendations on how to mitigate climate change, famously characterised climate change as: "...the greatest market failure the world has ever seen...." (Stern 2006: xviii). This economic framing of climate change is so widespread that we, by and large, have forgotten about it.

A consideration of how climate change has been framed as an economic issue strays into the territory of a substantial area of academic scholarship and policy debate on neoliberalism (a hugely important area of work, but one that is not directly covered in this book; see McCarthy and Prudham 2004; Harvey 2005). Neoliberalism is about "rolling back" the state and letting markets self-regulate, coupled with an antagonism towards state interference (McCarthy and Prudham 2004: 276): a type of governmentality that encompasses a shift in the role of the state away from regulation to managing markets (Mansfield 2007). Neoliberalism is not explicitly engaged with in Part II of this book (or elsewhere), because one of the objectives of *The Making of Low Carbon Economies* is to recognise and make visible the economic framing of climate change, but also to attempt to blur the boundaries of this economic framing by redefining what is 'economic'. What is being attempted here is thus a broadening out of the typically narrow contemporary definition of an economy to reflect what economies in practice actually involve (all sorts of things, as heterogeneous networks – see Part I), to embrace the possibility of diverse economies (Gibson-Graham 2008) and to remind us of the much more open definition of the economy that operated in the past. These are not issues that are central to debates about neoliberalism (McCarthy and Prudham 2004; Harvey 2005). Foucault's analysis is, however, relevant here because of his interest in exploring how economic subjects are framed, including the notion of the economy itself, and how this can change over time. As discussed in Chapter 1, it is in Foucault's 1978 lecture series entitled "Security, territory and population" (2007 translation) that he charts in detail a broadening in the meaning of economy and its scale of operation, from the family to the national and international scales. He explains how up until the seventeenth century 'economy' was about the sound management of the household – about caring and providing for the family unit: "...this idea of the economy ... at that time only ever referred to the management of a small *ensemble* comprising the family and the household" (Foucault 2007: 103). So it was not until relatively recently, in historical terms – the eighteenth century – that a new, broader notion of economy was introduced into political practice and, crucially, the economy became "a field of intervention for government" (2007: 95), as Foucault explains:

> To govern a state will thus mean the application of economy, the establishment of an economy, at the level of the state as a whole, that is to say, [exercising] supervision and control over its inhabitants, wealth, and the conduct of all and each, as attentive as that of a father's over his household and goods.
>
> (2007: 95)

In this way, Foucault's insights about how very different notions of the economy have existed in the past provide a platform to enable us to explore the contemporary issue of the reframing of existing economies, as well as the making of new economies and markets, in light of climate change. Foucault's account of the changing nature of the economy helps us understand the transition of climate change from a scientific problem to something that is part of the "...field of intervention for government" (Foucault 2007: 95). The essence of the climate change debate centres on how to introduce a "meticulous attention" (Foucault 2007: 95; see Chapter 5) to climate within (already existing) systems of economic management employed by the state.

A fourth insight provided by framing is that it is inherently political, because a key function of frames is to claim ownership (Rein and Schon 1993; Hajer 1995; Sharp and Richardson 2001). Particular groups of people involved in the policy process frame an issue in ways that make it understandable and solvable to them, fitting their beliefs, values, knowledge and professional expertise (Power 1991, 1997). This is clearly demonstrated in the chapters that comprise Part II of this book, where tensions between divergent framings of UK sustainable housing have led to conflict (Chapter 4); in contrast, very different framings of carbon accounting have meant that groups of people working in the field are barely aware of each other's existence (Chapter 5). Most often, framing leads to overt, public and highly political struggles for ownership between communities that have come together to frame the issue in their own way. Indeed, for Hajer and others (Litfin 1994; Hajer and Versteeg 2005), it is precisely this competition between frames that drives (policy) change, therefore making it a valuable field of study. By explicitly acknowledging and highlighting the making of diverse low carbon economies by different communities of practice and expertise – sometimes in opposition, sometimes happily co-existing – there are significant opportunities to encourage constructive dialogue, learning and policy change (Rein and Schon 1993).

Hajer describes the communities or networks of actors who coalesce around a particular frame as "discourse coalitions", defined as:

> [Coalitions who].... [d]evelop ... a particular way of talking and thinking ... These coalitions are unconventional in the sense that the actors have not necessarily met, let alone that they follow a carefully laid out and agreed upon strategy. What unites these coalitions and gives them their political power is that fact that its actors group around specific story-lines....
>
> (Hajer 1995: 13)

For Hajer, storylines are the 'glue' that unite discourse coalitions, defined as: "the essential discursive cement that creates communicative networks among actors with different or at best overlapping perceptions and understandings" (1995: 63). Chapter 4 on low carbon housing contrasts discourse coalitions with another type of community of actors – an advocacy coalition (Sabatier and Jenkins Smith 1993; Sabatier 1998), defined as: "people from a variety of positions . . . *who share a particular belief system* – i.e. a set of basic values, causal assumptions, and problem perceptions – and who show a non-trivial degree of co-ordinated [policy] activity over time" (Sabatier 1998: 115 [emphasis added]).

A fifth insight from scholars of framing is that it is inevitably incomplete. Framing is an active, ongoing process. As Callon observes specifically in relation to carbon markets:

> At the heart of markets we find debates, issues, feelings, matters of concern, dissatisfaction, regrets, and plans to alter existing rules, which cannot be internalised once and for all because they are linked to irreducible uncertainties, to what I have called framings which are never either definitive or unquestionable. This "hot" component of markets . . . always exists.
> (2009: 541)

There is thus a fragility to frames that is a helpful guide for analysis. For, instead of seeing (low carbon) markets and economies as stable and durable entities, framing is an approach that looks for the hidden work underway in maintaining an *illusion* of stability. Frames are inherently fragile. Scholarship that draws together STS ideas about the agency of materials and tangible things in providing durability and stability with the political science concept of framing is especially pertinent to thinking about the making of low carbon economies. This is demonstrated, for example, in Chapter 4, where it is explained how low carbon homes built by pioneers became part of 'mainstream' policy discourse.

Markets and economies, characterised as heterogeneous networks (see Part I), are seen as inherently fragile. Framing is thus similarly crucial to providing stability in this context, so market calculative processes are able to take place. In order to be part of a market, goods or commodities must be disentangled from their networks of connections and be stabilised within market frames – locked-in – so market actors can manage them in consistent, reliable ways. Callon (1998: 16) thus refers to framing as ". . . a process of disentanglement", also recognising the impossibility of enrolling and keeping market actors within frames, noting that: "Any frame is necessarily subject to overflowing" (1998: 18). Standardisation is, for example, explicitly identified by Çalışkan and Callon as a potential topic of research for 'pacifying goods' within markets: "Disentanglement is more stable . . . when a commodity has undergone specific processes of standardization that transform it into an entity described in both abstract and precise terms, certified and guaranteed by a series of textual and material devices" (Çalışkan and

Callon 2010: 8). These ideas about calculation and standards and their integral role in making markets and economies have links to the theme of Part III of the book on commensuration.

Sixth, attention to the fragility of frames highlights the possibility of significant change. While Callon and others working in the performativist school of economic sociology concentrate on the ways in which individual frames deconstruct themselves via "overflows", in studies of the policy process analytical attention is focused more directly on the interactions and conflicts *between* frames. Rein and Schon (1993), for example, distinguish between policy disagreements and policy controversies: disagreements occur within the same frame, controversies are between different frames, and thus are much more difficult to resolve. Significant shifts in dominant frames is also a topic of enquiry central to transitions theory, with its interest in how, when and why radical innovations diffuse – i.e. in explaining significant moments of change (see the introduction to Part I). However, societal expectations regarding change and the complexity of policy processes have been mostly neglected by transitions theory scholars, who are typically more focused on the technologies themselves (especially in the early work on transitions; see, for example, Kemp *et al.* 1998; Geels 2004). Transitions scholarship takes as its starting point that it is possible to intervene in sociotechnical systems and change the direction of system growth (see, for example, the work on 'strategic niche management'; see Schot *et al.* 1994; Kemp *et al.* 1998), ignoring how social actors come together and might, for instance, dispute the governance of sociotechnical transitions or have conflicting views about the trajectory of change (Meadowcroft 2009; Murphy and Smith 2013). Because of this, transition approaches have been criticised as overly 'functionalist', without due regard to issues of agency, politics and discourse (Meadowcroft 2009).

There are a small number of transition scholars, however, who have argued for a more central role for discourse and framing in conceptualising large-scale processes of change and innovation. For example, Smith *et al.* (2005: 1495) argue that: "The process of articulating knowledge of environmental pressures is typically as essential as processes of active governance in the effective realisation of change in socio-technical regimes". For Smith *et al.* (2005), pressures to change within sociotechnical systems ('selection pressures') exist all the time, they are constant. What is crucial, therefore, in seeking to locate and understand change, is how periodically these pressures are articulated in such a way that they mobilise change. In other words, the form of this articulation – how the issues are framed – matters: for instance, whether pressures are framed so as to be orientated in the same direction, and whether they are explicit and in a form that enables response by the sociotechnical system. Smith *et al.* (2005) thus provide a key starting point for thinking about how ideas concerning discourse and framing might be integrated more centrally within transitions scholarship. They propose a 'quasi-evolutionary model' of sociotechnical regime change, whereby change is a function of two processes: selection pressures on the regime (where particular framings of a problem increase or decrease these pressures) and the resources available to adapt to them.

Others have also, in different ways, considered the role of discourse and framing within transitions (Smith and Kern 2007; Lovell 2008). One novel and quite different way in which framing becomes relevant for transitions is in explaining the discursive appeal of the theory of sociotechnical transitions itself – a theory that has become very popular in recent years, amongst both academics and sustainability practitioners. Smith and Kern (2007) use such an approach to examine the rise of transitions discourse in Dutch sustainability policy, explaining the discursive appeal of its relatively straightforward pathway for change and its consequent traction within government. Thus, the concept of sociotechnical transitions has itself become a powerful way of framing low carbon policy and technology debate in countries such as the Netherlands and, to a more limited extent, the UK (see, for example, HM Government 2009).

In summary, in this short introduction to Part II, I have explained what framing means and how it has been applied as a concept across several different disciplines. Frames are a lens on the way that problems are viewed, discussed and resolved and are thus are of vital importance in understanding how climate change, as a relatively new and complex problem, has come to be understood and dealt with – i.e. in exploring how and why particular solutions have emerged. Chapters 4 and 5 examine two examples of this process: in UK housing (Chapter 4) and carbon accounting (Chapter 5).

Bibliography

Abolafia, M. Y. (1998). "Markets as cultures: an ethnographic approach". *The Laws of the Markets*. M. Callon (ed.). Oxford, Blackwell Publishers/The Sociological Review: 69–85.

Barry, A. and D. Slater (2007). "Introduction". *The Technological Economy*. A. Barry and D. Slater (eds). London and New York, NY, Routledge: 1–27.

Bickerstaff, K., I. Lorenzoni, N. F. Pidgeon, W. Poortinga and P. Simmons (2007). "Reframing nuclear power in the UK energy debate: nuclear power, climate change mitigation and radioactive waste". *Public Understanding of Science* **15**: 1–25.

Boyd, E., M. Boykoff and P. Newell (2011). "The 'new' carbon economy: what's new?" *Antipode* **43**(3): 601–611.

Çalışkan, K. and M. Callon (2010). "Economization, part 2: a research programme for the study of markets". *Economy and Society* **39**(1): 1–32.

Callon, M. (ed.) (1998). *The Laws of the Markets*. Oxford, Blackwell Publishers/The Sociological Review.

Callon, M. (2007). "What does it mean to say that economics is performative?" *Do Economists Make Markets? On the Performativity of Economics*. D. MacKenzie, F. Muniesa and L. Siu (eds). Princeton, NJ, Princeton University Press: 311–357.

Callon, M. (2009). "Civilizing markets: carbon trading between *in vitro* and *in vivo* experiments". *Accounting, Organizations and Society* **34**(3–4): 535–548.
Dean, M. (1999). *Governmentality: Power and Rule in Modern Society*. London, Sage Publications Ltd.
Dryzek, J. (1997). *The Politics of the Earth: Environmental Discourses*. Oxford, Oxford University Press.
Etzion, D. and F. Ferraro (2010). "The role of analogy in the institutionalization of sustainability reporting". *Organization Science* **21**(5): 1092–1107.
Foucault, M. (2007). *Security, Territory, Population*. Basingstoke, Palgrave Macmillian.
Geels, F. W. (2004). *Technological Transitions and System Innovations: A Co-Evolutionary and Socio-Technical Analysis*. Cheltenham, Edward Elgar.
Gibson-Graham, J.-K. (2008). "Diverse economies: performative practices for other worlds". *Progress in Human Geography* **32**(5): 613–632.
Hajer, M. and W. Versteeg (2005). "A decade of discourse analysis of environmental politics: achievements, challenges, perspectives". *Journal of Environmental Policy and Planning* **7**(3): 175–184.
Hajer, M. A. (1995). *The Politics of Environmental Discourse: Ecological Modernisation and the Policy Process*. Oxford, Clarendon Press.
Hajer, M. A. and H. Wagenaar (eds) (2003). *Deliberative Policy Analysis: Understanding Governance in the Network Society*. Cambridge, Cambridge University Press.
Harvey, D. (2005). *A Brief History of Neoliberalism*. Oxford, Oxford University Press.
Hastings, A. (1999). "Analysing power relations in partnerships: is there a role for discourse analysis?" *Urban Studies* **36**(1): 91–106.
Her Majesty's (HM) Government (2009). "The UK Low Carbon Transition Plan: national strategy for climate and energy". London, UK Government.
Hoffman, A. J. and M. J. Ventresca (1999). "The institutional framing of policy debates: economics versus the environment". *American Behavioural Scientist* **42**(8): 1368–1392.
Jacobs, K., J. Kemeny and T. Manzi (2003). "Power, discursive space and institutional practices in the construction of housing problems". *Housing Studies* **18**(4): 429–446.
Kemp, R., J. W. Schot and R. Hoogma (1998). "Regime shifts to sustainability through processes of niche formation: the approach of strategic niche management". *Technology Analysis and Strategic Management* **10**(2): 175–195.
Laws, D. and M. Rein (2003). "Reframing practice". *Deliberative Policy Analysis: Understanding Governance in the Network Society*. M. A. Hajer and H. Wagenaar (eds). Cambridge, Cambridge University Press: 172–206.
Litfin, K. T. (1994). *Ozone Discourses: Science and Politics in Global Environmental Cooperation*. New York, NY, Columbia University Press.
Lohmann, L. (2005). "Marketing and making carbon dumps: commodification, calculation and counterfactuals in climate change mitigation". *Science as Culture* **14**(3): 203–235.
Lohmann, L. (2009). "Toward a different debate in environmental accounting: the cases of carbon and cost-benefit". *Accounting, Organizations and Society* **34**(3): 499–534.
Lövbrand, E. and J. Stripple (2011). "Making climate change governable: accounting for carbon as sinks, credits and personal budgets". *Critical Policy Studies* **5**(2): 187–200.
Lovell, H. (2004). "Framing sustainable housing as a solution to climate change". *Journal of Environmental Policy and Planning* **6**(1): 35–56.
Lovell, H. (2008). "Discourse and innovation journeys: the case of low energy housing in the UK". *Technology Analysis & Strategic Management* **20**(5): 613–632.

Mansfield, B. (2007). "Articulation between neoliberal and state-orientated environmental regulation: fisheries privatisation and endangered species protection". *Environment and Planning A* **39**: 1926–1942.

McCarthy, J. and S. Prudham (2004). "Neoliberal nature and the nature of neoliberalism". *Geoforum* **35**: 275–283.

Meadowcroft, J. (2009). "What about the politics? Sustainable development, transition management, and long term energy transitions". *Policy Sciences* **42**(4): 323–340.

Murphy, J. and A. Smith (2013). "Understanding transition–periphery dynamics: renewable energy in the Highlands and Islands of Scotland". *Environment and Planning A* **45**(3): 691–709.

Rein, M. and D. Schon (1993). "Reframing policy discourse". *The Argumentative Turn in Policy Analysis and Planning*. F. Fischer and J. Forester (eds). London, UCL Press Ltd.: 145–166.

Sabatier, P. A. (1998). "The advocacy coalition framework: revisions and relevance for Europe". *Journal of European Public Policy* **5**(1): 98–130.

Sabatier, P. A. and H. C. Jenkins Smith (eds) (1993). *Policy Change and Learning: An Advocacy Coalition Approach*. Boulder, CO, Westview Press.

Schot, J. W., R. Hoogma and B. Elzen (1994). "Strategies for shifting technological systems: the case of the automobile system". *Futures* **26**(10): 1060–1076.

Sharp, L. and T. Richardson (2001). "Reflections on Foucauldian discourse analysis in planning and environmental policy research". *Journal of Environmental Policy and Planning* **3**(3): 193–209.

Smith, A. and F. Kern (2007). "The transitions discourse in the ecological modernisation of the Netherlands". SPRU Working Paper No. 160. Brighton, SPRU, University of Sussex.

Smith, A., A. Stirling and F. Berkhout (2005). "The governance of sustainable sociotechnical transitions". *Research Policy* **34**: 1491–1510.

Stern, S. N. (2006). "The Stern Review on the economics of climate change". London, HM Treasury.

4

LOW CARBON HOUSING IN THE UK

Key findings

- Low carbon housing in the UK has been framed, or reframed, from already existing sustainable housing.
- Different organisations and communities have been involved in this reframing (characterised here as a discourse coalition; see Hajer 1995), united by a shared way of thinking and talking about climate change.
- An advocacy coalition (a group united by shared values; see Sabatier and Jenkins Smith 1993) best characterises those involved in building and developing the long-standing sustainable housing.
- Climate change was just one of the many policy problems that was jostling for attention on the UK Government's housing policy agenda in the late 1990s – low carbon economies are inevitably a compromise and typically incorporate other policy objectives as well.
- The chapter provides an example of how new low carbon economies (in this case, housing) are framed out of existing practices and ways of doing.
- This reframing of infrastructures and materials can happen quickly – i.e. the obsolescence of infrastructures does not necessarily have to take place, for there are opportunities to rebrand existing things and practices as low carbon.
- The material presence of low carbon housing built in the UK in the late 1990s and early 2000s (i.e. the fact that it was there, in place) proved vital in giving the UK Government the confidence to introduce a radical new policy in 2006 – namely, that all new housing built in England and Wales must be zero carbon from the year 2016.
- Existing sustainable and zero carbon homes did not provide a solid evidence base for the introduction of the 2016 zero carbon homes target, but rather played a pivotal *discursive* role in the construction of a new zero carbon UK house building sector. It is a finding that problematises the objective, experimental role of innovation niches within transitions theory.

Introduction

This chapter examines a formative period in the making of low carbon economies for housing: the late 1990s and early 2000s. It thereby explores the foundations for a radical UK Government policy decision taken in 2006 to mandate that all new housing in England and Wales must be zero carbon by the year 2016 (DCLG 2006). I was undertaking research (during the period 2001–2003), at a time when UK Government interest in low carbon housing was rapidly rising. I approached several key organisations involved in a small number of high profile, sustainable, ultra-low energy homes that had been built during the late 1990s, enquiring if I could conduct fieldwork with them. I was told that demand had been so high from researchers that it was not possible for me to do this, but that I could, however, sign up to take a tour around the developments. Thinking this was a second-best option, I rather reluctantly did so. However, when I arrived at the two developments I was studying who were offering tours ('BedZed', located in South London, and the Hockerton Housing Development in the East Midlands), I was in for a surprise: on each of the tours there were present directors of major private sector house builders, as well as senior UK civil servants. They were there to learn, and it was an extremely useful opportunity for me to listen to what they were asking about and to have a chance to talk to them myself. I learnt from observing them and talking to others that part of what they were interested in was how they could use these existing sustainable housing developments – one of which was extremely eye-catching and photogenic – to promote policies and initiatives that they were soon to announce. For example, Patricia Hewitt, the then UK Secretary for Trade and Industry, used BedZed as a location to announce a new government solar power initiative (DTI 2002). Similarly, the Liberal Democrat Party Leader also visited BedZed that year, because he "was making an environment announcement later that day and wanted a photo to go with any publicity" (Email correspondence with BedZed Communications Officer, 25 October 2004).

The close government attention to existing sustainable housing as a basis for promoting new low carbon policies continued. This chapter explores how the UK Government policy for all new homes in England and Wales to be zero carbon by 2016 has its roots in the turn of the century reframing of existing sustainable housing as 'low carbon'. The analysis concentrates principally on how and why this reframing occurred, based on extensive fieldwork undertaken over the period 2001–2003.[1] As is shown, climate change was jostling amongst a number of other policy problems high on the UK Government's policy agenda at the time, including modernisation of the construction industry and meeting the demand for new housing. In the second half of this chapter, I reflect, with the benefit of the passage of time, on how well the empirical findings fit with transitions theory ideas about patterns of change in large infrastructure systems – in particular, how niches are mainstreamed and the role of framing therein.

A key reason behind the growing interest in reframing sustainable housing as 'low carbon' during the late 1990s and early 2000s was the increasing importance of climate change as a policy issue in the UK. It was during the 1990s that climate change first became an important policy problem within the UK, along with other industrialised countries (Newell 2000; Bulkeley and Betsill 2003; Lovell *et al.* 2009). During this period, the UK was one of the most proactive countries in the international climate change negotiations (Grubb 2002) and, for example, set a carbon dioxide reduction goal of 20 per cent below 1990 levels by 2008–2012, thus going significantly beyond its Kyoto Protocol target of 12.5 per cent greenhouse gas reduction on 1990 levels (DETR 2000). "Climate change: the UK programme", published in the year 2000, outlined in detail for the first time how the UK's greenhouse gas emissions would be reduced (DETR 2000). The UK Government accepted recommendations made by the Royal Commission on Environmental Pollution (RCEP) in its influential energy report "Energy – the changing climate" (RCEP 2000) to reduce greenhouse gas emissions by 60 per cent by 2050 (DTI 2003a). This then paved the way for the Climate Change Act 2008, which deepened the UK Government's commitment to greenhouse gas emission reduction to 80 per cent by 2050 (HM Government 2008). In short, there was a lot of climate change policy activity during the 1990s and early 2000s – in particular, ambitious target setting – which left many in government, and beyond, asking how these targets would actually be met: the search for ready-made low carbon solutions began.

Sustainable housing is defined here as housing that has environmental and social benefits above those of an average new UK house ('average' here being the standard defined by the UK Building Regulations, at the time the "Approved document L1 – The Building Regulations 2000 (Conservation of fuel and power in dwellings)" (see DTLR 2002)). This broad definition is used because much of the discussion in this chapter is about struggles over the definition of sustainable housing: there is no consensus on a more specific definition. Low carbon housing is defined as housing which has lower greenhouse gas emissions (principally carbon dioxide) compared with an average new house built to the UK Building Regulations (which equated, at the turn of the century, to modelled emissions of less than one tonne of carbon per year; see DTLR 2002). A low carbon home typically incorporates one or more of the following features: passive low energy design, a thermally efficient built form and use of renewable energy technologies.

This chapter is structured as follows. First, the sustainable housing advocacy coalition is introduced. Second, the different types of policy problems that sustainable housing was linked to during the 1990s are outlined, and the framing of sustainable housing as low carbon housing – the winner in this policy jostling – is explained, drawing on analysis of two common, discursive "storylines" (Hajer 1995) used by those interested in promoting low carbon housing. Third, reflections are made on the pivotal role of already existing low and zero carbon homes in the development of the 2016 zero carbon homes policy, drawing on transitions theory. In conclusion, how low carbon economies are made from a selective reframing of existing economies and ways of doing is discussed. The process of making low carbon economies in existing areas of activity and policy (housing) is therefore inevitably about stitching together the old and the new; taking some ideas, technologies and institutions and discarding others. It is a selective process, and, seen in this way, it is evident that the making of low carbon economies is as much a social and political project as a technical one.

The sustainable housing advocacy coalition

The UK's contemporary sustainable housing movement emerged in the early 1970s (*The Ecologist* 1972; Bhatti *et al.* 1994; Barton 1998; Smith *et al.* 1998; Chappells and Shove 2000), concurrent with an increased public awareness of environmental issues and an upsurge in radical deep green environmentalism (Sandbach 1980; Weale 1992; Porter and Brown 1996; Dryzek 1997). Examples of sustainable housing developments from this period include the Centre for Alternative Technology (CAT) in Wales and the Findhorn Ecovillage in Scotland. Those involved in sustainable housing from the 1970s onwards are best characterised as an advocacy coalition (Sabatier and Jenkins Smith 1993; Sabatier 1998), defined as: "people from a variety of positions . . . *who share a particular belief system* – i.e. a set of basic values, causal assumptions, and problem perceptions – and who show a non-trivial degree of co-ordinated [policy] activity over time" (Sabatier 1998: 115 [emphasis added]). Members of the sustainable housing advocacy coalition share so-called 'deep green' environmental values and beliefs: they believe radical societal changes are necessary in order to achieve environmental sustainability, such as governance via small-scale self-sufficient communities (Dobson 2000). Further, deep greens believe use of more efficient technologies and reform of existing institutions (the 'light green' approach) will not be sufficient to solve environmental problems: a fundamental shift in attitudes and consciousness is required (Dobson 2000).

The key distinction between an advocacy coalition and other policy categorisations, including discourse coalitions, is that members are united by these shared values and beliefs. In other words, the 'glue' that binds those involved in sustainable housing is their deep green environmental values. The sustainable housing advocacy coalition can, however, only loosely be defined as a type of *policy* community, because those involved believe in small-scale community action outside of the state. In other words, sustainable housing advocates active in the UK since the 1970s have intentionally distanced themselves from public sector organisations and have not

Table 4.1 Belief system of the sustainable housing advocacy coalition

Deep core *Fundamental normative and ontological axioms*	*Policy core* *Fundamental policy positions concerning the basic strategies for achieving core values within the subsystem*	*Secondary beliefs* *Instrumental decisions and information searches necessary to implement policy core*
Environment prioritised: ecocentric or deep green values and beliefs.	Live according to values: practice sustainable lifestyles.	Autonomous provision of resources (energy, water, food. etc.).
Anti-capitalist – environmental problems indicate a fundamental flaw with modern Western lifestyle.	Holistic approach to sustainability: economic, social and environmental issues all important.	Engagement with government not a priority.
Future generations and nature of greatest concern.	Government is cause of the problem: progress is via small-scale self-governing communities.	Reduce demand for resources before considering (sustainable) supply options.

Table 4.2 Principle founding actors of the UK sustainable housing advocacy coalition

Name of actor	*Brief description*
Centre for Alternative Technology (CAT), Machynlleth, Wales 1973+	A community development in an old quarry site in rural Wales, established in 1973. It is a self-built autonomous development (energy and water self-sufficient). It also operates as a sustainable housing education and resource centre and runs residential courses (see CAT 1995). CAT also publishes a quarterly sustainability magazine called *Clean Slate*.
Findhorn Ecovillage, Scotland 1962+	A self-built community in rural Scotland. It was established in 1962, and building on site started in the early 1970s. As with CAT, there is an education centre and residential courses (see Findhorn Ecovillage 2003).
Communes Network and Diggers and Dreamers 1968+	The Communes Network started as the Communes Movement in 1968, founded by the Selene Community in Wales. Amongst other activities, it produced a bimonthly magazine *The Communes Journal*, which had a print run of 3,000 copies. In 1975, it became the 'Communes Network', a more loosely connected organisation, which still operates informally today. Some members of the Communes Network have formed 'Diggers and Dreamers' – a self-build community organisation that aims to help self-builders to network and to access information on self-build housing (Dawling 1992).
Undercurrents magazine 1972 – early 1980s	*Undercurrents* was a radical environmental magazine published in the 1970s and early 1980s (commenced in 1972) and was regarded as the alternative movement's journal during this period. Its subtitle was "the magazine of radical science and people's technology". It focused, in particular, on sustainable housing communities active during the 1970s.

been primarily focused on effecting policy change. Their solutions are based on small-scale, self-sufficient and self-governing communities (Wood 1990; European Eco-village Network 2003). The deep core, policy core and secondary beliefs (after Sabatier 1998) of the sustainable housing advocacy coalition are outlined in Table 4.1. Table 4.2 briefly describes some of the principal actors involved in the UK sustainable housing advocacy coalition.

The following section examines how and why sustainable housing was actively framed as a solution to a range of policy problems during the 1990s – a process driven mainly by the UK Government. The framing of sustainable housing as a solution to climate change is then analysed in further detail.

Framing sustainable housing as a solution

As noted in the introduction to Part II, framing is important because it sets the boundaries around a policy issue and allows ownership of it by certain actors.

Further, framing inherently involves the framing of solutions, as well as problems. Solution framing is, of course, linked to problem framing, as how one categorises an issue as a problem necessarily sets parameters on the solutions that are sought. However, the relationship between solutions and problems is not usually linear, but fluid, with events and politics having a strong influence (Kingdon 2003). Thus, there are situations in which existing policies or activities are reframed as a solution to a different, often new, policy problem. Solution framing is described by Kingdon (2003) in his 'garbage can' model of the policy process, whereby "streams" of problems, policies and politics co-exist, sometimes merging to form a coherent policy "package": "...people in and around Government sometimes do not solve problems. Instead, they become advocates for solutions and look for current problems to which to attach their pet solution" (2003: 123).

Sustainable housing in the UK during the 1990s and early 2000s was framed as a solution to a number of policy problems (see Table 4.3), ranging from modernisation of the construction industry to meeting the demand for new housing. There was hence a struggle for ownership of sustainable housing across a range of government departments and institutions and between different types of housing provider (social, private sector and self-build). These different framings of what the policy problem was with respect to housing co-existed, and climate change jostled amongst them, ultimately emerging as the overriding issue (Lovell 2004). The suitability of sustainable housing as a solution to any particular problem depends, of course, on how sustainable housing is defined. The intensity of the debate over the meaning of sustainable housing during this period provides evidence of this struggle (see, for example, ENDS 2001; ERM 2002; Rudlin 2002; TCPA 2002).

The newer, more mainstream, actors that became active in the sustainable housing sector in the 1990s were mostly united through shared discourse rather than values, and thus Hajer's (1995) notion of a discourse coalition best characterises them (see the introduction to Part II for a fuller explanation of the term). Hajer defines discourse coalitions as groups of actors active in the policy process who "... develop ... a particular way of talking and thinking ... What unites these coalitions and gives them their political power is that fact that its actors group around specific story-lines..." (Hajer 1995: 13). For Hajer, "storylines" are the 'glue' that unite discourse coalitions, defined as: "the essential discursive cement that creates communicative networks among actors with different or at best overlapping perceptions and understandings" (1995: 63). The identification of a number of discourse coalitions actively competing within housing policy, presenting sustainable housing as a solution to several different policy problems (see Table 4.3), is in keeping with the conceptualisation of multiple discourse coalitions active in any one policy subsystem (Hajer 1995; Dryzek 1997).

Table 4.3 The framing of sustainable housing as a solution to UK policy problems at the start of the twenty-first century.

Policy problem sustainable housing is being framed as a solution to	*Organisations involved in framing*	*Examples of policies, programmes and housing developments*
Meeting the demand for new housing (*c.*3 million new homes needed by 2016)	Office of the Deputy Prime Minister (ODPM); private sector house builders; English Partnerships; Rethinking Construction (the Housing Forum); local authority planners; the Town & Country Planning Association (TCPA); WWF.	Sustainable Communities Plan (ODPM 2003b). Millennium Communities Programme (English Partnerships 2003). WWF One Million Sustainable Homes (OMSH) campaign (WWF 2003). Housing developments: • Plans for a 'Zed squared' zero energy, zero waste development in the Thames Gateway (Desai 2003); • West Stevenage development (The West Stevenage Consortium 2002); • Ashton Green, Leicester.
Lack of innovation in the construction industry	Rethinking Construction (including the Housing Forum); ODPM; DTI.	"Six guiding principles to improve the sustainability of the housing construction industry" (The Housing Forum 2002). Speech by Construction Minister Brian Wilson April 2003 – green housing and housing sector modernisation (DTI 2003b). Off-site manufacture report – "Homing in on excellence" (The Housing Forum 2001). Housing developments: • Greenwich Millennium Village (English Partnerships 2003); • Social housing – INTEGER projects at Maidenhead and Sandwell (Ekins 2002).
Poor quality of existing housing stock	ODPM; The Housing Corporation; Registered Social Landlords (RSLs); local authorities; Regional Housing Boards.	Sustainable Communities 'Pathfinder' Regeneration Areas (ODPM 2003b). Decent Homes (ODPM 2003a). Urban regeneration companies Housing developments: • Inner-city tower block refurbishment – e.g. Glastonbury House, Pimlico, London (Brown 2003); • Plans for 7,000 new and refurbished 'ultra green' homes in Oldham (Tickle 2003).

Fuel poverty	Local authorities; Energy Efficiency Advice Centres; Energy Saving Trust (EST); energy utilities; Association for the Conservation of Energy (ACE).	UK Fuel Poverty Strategy (DEFRA 2001). The Energy White Paper (DTI 2003a). Warm Homes programme and the Home Energy Conservation Act (HECA) (DEFRA 2000b). Housing developments: • Boughton Energy Village, Newark and Sherwood District Council, East Midlands; • Ravenscliffe, Bradford, North British Housing Association (1999), 64 low energy timber frame houses.
Traffic congestion	Department for Transport; local authority planners; Transport 2000.	Transport 10-year plan (DEFRA 2000a). Housing developments: • Slateford Green, Edinburgh; • BedZed, south London.
Meeting renewable energy generation targets	Energy utilities; Department for Trade and Industry (DTI); renewable energy companies; The Countryside Agency; Building Research Establishment (BRE); RSLs.	"Clear Skies" community and household grant programme (BRE 2003). Generating Solar Homes (Generating Solar Homes 2002). The Energy White Paper (DTI 2003a). Housing developments: • North Nines, Edmonton, London (Laing Homes and Solar Century); • Sherwood Energy Village, near Ollerton, East Midlands.
Climate change mitigation	EST; Department for Environment, Fisheries and Rural Affairs (DEFRA); The Carbon Trust; local authorities; private sector house builders; RSLs; green architects; TCPA; WWF; DTI; BRE.	Energy White Paper (DTI 2003a). (RCEP 2000) "Energy – The Changing Climate, 22nd report." Housing Developments: • The Vales' Autonomous House; • Hockerton Housing Development, Notts.; • BedZed, South London; • Greenwich Millennium Village.

There are two explanations as to why sustainable housing was targeted by discourse coalitions for solution framing. First, there are push factors – characteristics of sustainable housing itself that made it an attractive proposition – and second, pull factors – motivations for the framers to claim ownership. The first push factor is that a reasonably large amount of sustainable housing already existed in the 1990s. As noted, building sustainable housing developments was the main political activity of the sustainable housing advocacy coalition. The material presence of these sustainable housing developments rendered them particularly liable to discursive capture and rebranding. The existing sustainable homes were often photogenic and therefore media friendly (see, for example, the Hockerton Housing Project 2003). The second push factor stemmed from sustainable housing having emerged from outside of government – a 'bottom-up' social movement. It was thus relatively free of associations with past policy programmes and therefore easier to frame as a solution.

The pull factors, which explain why sustainable housing was of interest to policymakers and other mainstream institutions, are, first, that it was potentially a solution to some persistent policy problems, such as fuel poverty (see Table 4.3). Second, because some sustainable housing already existed, it was relatively easy to give the impression that that rapid progress had already been made on an issue. This was particularly important with climate change, because at the time, as noted, there was growing concern about how ambitious UK Government targets would be met (RCEP 2000; SDC 2003). Further, the existence of these homes demonstrated that new sustainable technologies worked and thus reduced the risk for more mainstream actors who wished to use them. The material reality of these homes grounded rhetoric or discourse and thus lent the low carbon discourse coalition credibility.

Climate change was the problem to which existing sustainable housing, in the end, became most closely associated, through the work of an active low carbon discourse coalition. Through their reframing of sustainable housing as 'low carbon', a new type of low carbon economy was made, bringing together elements of the old and the new. Existing sustainable housing was discursively remade in response to a new growing policy problem: climate change. Further, existing sustainable housing was reframed as something that was cost-effective and could be of financial benefit to the UK house building industry (Lovell 2004). An important element in this process of solution framing of low carbon housing was the creation of distance between the discourse coalition and the deep green values of the sustainable housing advocacy coalition, through the use of ecological modernist discourse. There are several definitions of ecological modernisation (see Mol and Spaargaren 2000 for an overview); here, ecological modernism is defined as a policy discourse, described as: "the discourse that recognises the structural character of the environmental problematique, but none the less assumes that existing political, economic, and social institutions can internalise the care for the environment" (Hajer 1995: 25). Further, the ecological modernist policy discourse used by the low carbon discourse coalition constitutes a weak version of ecological modernisation, in that eco-efficiency and technology solutions dominate (Christoff 2000). It thus stands in strong contrast to the deep green values and beliefs of the sustainable housing advocacy coalition. In this way, the reframing undertaken by the low carbon discourse coalition was in large part about making existing (deep green) sustainable housing work within a more mainstream, profit-seeking house building economy. As is explained further later, the emphasis of the low carbon discourse coalition was primarily on economic feasibility. In this way, they 'cherry-picked' certain ideas and technical features of existing sustainable housing and sought to bring them into mainstream policy debates and economies through the use of particular low carbon storylines, to which I now turn.

Low carbon storylines

Storylines are defined by Hajer (1995: 63) as the 'glue' or "discursive cement" that unites discourse coalitions (see earlier discussion and the introduction to Part II). Two discursive storylines that helped unite the diverse low carbon discourse coalition are about housing 'life cycles' and 'smart housing'. In this section the storylines are explained and the responses they provoked from the sustainable housing advocacy coalition are examined. Both the housing life cycle and smart housing storylines are about making low carbon housing economic – i.e. about remaking sustainable housing and positioning it as something that can function well within a profit-seeking capitalist economy. These storylines thus involve a *re*framing of existing sustainable housing according to quite different metrics. The response from the sustainable housing advocacy coalition was to try to widen the discourse from this narrow, economically orientated low carbon framing, and tensions are evident between these two coalitions.

The life cycle storyline

In the housing sector a 'life cycle' approach refers to the practice of examining economic (and environmental) costs and benefits over the lifetime of a house or housing development. In other words, a life cycle approach takes a long-term view as to when initial investment capital may be recouped. Life cycle storylines are used in response to the (often posed) question: 'Does sustainable housing cost more?' about which there remains much confusion. This is because the answer depends on several factors: the time frame of consideration; the type of housing (i.e. private or social sector or self-build); and whether it is newbuild housing or refurbishment of old housing (and if so, how old the housing is).

The life cycle storyline was used extensively by the low carbon discourse coalition to demonstrate the economic credentials of sustainable and low carbon housing. The life cycle was invoked as a familiar metaphor, including terms such as 'payback periods', and the economic discourse of costs and benefits and cost-effectiveness, designed to reassure government and the house building industry that sustainable and low carbon housing was feasible to integrate into the mainstream house building economy. The storyline proceeds as follows: it is sensible to invest extra money at the design and construction stages of a house, as it can be recouped when the house is sold, because there is significant consumer demand for low carbon, better quality housing, and it can therefore be sold at a premium. Further, better quality housing leads to lower (or even non-existent) utility bills, thus over the lifetime of the house these costs are recouped by the householder. Thus, for example, a local authority manager responsible for planning a large low carbon housing development near Leicester used this type of life cycle storyline to justify the council selling the land to a developer at a lower price, and their intention to retain a financial stake in the development:

> So the houses are more expensive because they are more popular, because it is a super place to live, well designed very attractive, low energy.... There will be a premium on the house prices so the [local] authority will get some of that back through each house that is sold off. So ultimately it will be self-financing, but there is a cost up front through the land sale.
>
> (Interview, December 2002)

The economic life cycle storyline is also used in part as a way to remove environmental values from the debate and thus distance low carbon housing actors from the sustainable housing advocacy coalition. In other words, low carbon housing is portrayed primarily as a sensible financial investment, and in this way the storyline recasts the original ambitions and aims of the sustainable housing advocacy coalition.

The response of the sustainable housing advocacy coalition to the life cycle storyline was to counter the idea that a financial profit must be made in order for an investment decision to be deemed rational. In other words, they disputed that cost-effectiveness should be the sole criteria by which sustainable and low carbon housing is evaluated; the payback of any investment, in their view, must include environmental and social criteria. For example, a local authority manager involved in planning a sustainable housing development describes a situation of environmental benefit through higher capital investment: "So the [local] authority is ultimately having to pay for the energy efficiency approach. But we accept that the pay back to the environment is worth it" (Interview, December 2002).

Another example is the decision-making by the owners and builders of the UK's first modern autonomous house (i.e. not connected to the utility grid) (Vale and Vale 1980). The Vales decided to purchase expensive photovoltaic (PV) panels for their house, essentially a non-economically rational decision, as at the time the financial savings in annual energy bills were small (there was no government subsidy) and the payback period was approximately fifty years. The story told about this decision by members of the sustainable housing advocacy coalition is as follows, here told by a local authority energy manager:[2]

> . . . the Vales' response to how you can justify spending £15,000 on (PVs) . . . for a £150 [per annum] saving was beautiful. . . . It is normal to have a £20,000 kitchen in a high status house isn't it? Where is the pay back in a £20,000 kitchen . . .? "So I look out the window at my PVs and it gives me great pleasure. £20,000 kitchen annoys me, well what a waste of money." It's down to values isn't it?
>
> (Interview, August 2002)

The Vales were thereby contesting the idea that financial payback is the only consideration in a rational purchase decision. The payback for them instead comprised less tangible, non-monetary returns, such as reducing their consumption of non-renewable resources: a lifestyle decision in keeping with their deep green values.

The smart housing storyline

A second storyline that was commonly used by the low carbon housing discourse coalition centres on the notion of 'smart housing'. Smart housing can be narrowly defined as that which "use[s] electronic networking technology to integrate the various devices and appliances found in almost all homes . . . so that the entire home can be controlled centrally – or remotely – as a single machine" (Pragnell *et al.* 2000). However, in this (sustainable and low carbon) context, smart housing was used at the time to denote housing in which householders were not required to modify their behaviour in order to become less resource intensive. In other words, one can live in a smart house and continue to behave as normal. Whilst at first sight the smart housing storyline is not directly about making low carbon economies, the high-tech futuristic orientation of this storyline captures a particular way of rethinking existing sustainable housing as innovative and therefore marketable. The smart housing storyline is technology focused, with a deliberate emphasis on the lack of requirement for social or institutional change: one does not have to adopt the deep green values of the sustainable housing advocacy coalition in order to live more sustainably. Thus, for example, a sustainable housing manager at an environmental charity involved in building sustainable housing developments described their approach in smart housing terms:

> So what we're trying to do on our [housing] developments is, it's all in there, you buy the house and it's there, you don't have to think about it, you're not even aware of it. But actually when your water bill comes through its only £50 because you've got a 2 litre flush toilet, and you've got low pressure aerated taps.... And they are all put in in a way that you would be nuts to want to replace them with something else.
>
> (Interview, June 2002)

Similarly, a sustainability manager in the social housing sector explained: "We try to promote passive [technologies], so that householders don't even need to know that they're making an environmental saving" (Interview, June 2002). The chief architect at the well-publicised BedZed sustainable housing development in South London, completed in 2001, also discussed the approach of the BedZed team using smart housing ideas:

> [We're] trying to come up with a lifestyle that makes it easier and more convenient to live a lower impact existence, than by using conventional alternatives. So what we're saying is that *if you're prepared to work with the infrastructure we've provided*, you can achieve really quite astonishing things. It is possible to live [at BedZed] and be pretty close to carbon neutrality.
>
> (Bill Dunster, quoted in Lowenstein 2001: 16 [emphasis added])

The response of the sustainable housing advocacy coalition to the smart housing storyline was that such approaches may go some way to reducing resource consumption, but that ultimately some modification of householder behaviour and greater householder awareness and education is required. Smart housing posed a challenge for sustainable housing advocates, as it was a direct attempt to prove that environmental values are not a necessary component of successful sustainable, or low carbon, homes. Adding smart technologies to a home was challenged by sustainable housing advocates as an 'end of pipe' solution: a short-term technological fix to a problem which requires institutional and social change (see, for example, CAT 1995; Liddell and Grant 2002). The sustainable housing advocacy coalition stressed that without the ideological commitment of the householder (i.e. deep green values), and an associated level of knowledge and motivation, the smart technology in the houses would simply not function properly. A sustainable housing manager at a UK charity, for instance, described the difficulty of making the technologies work in their holiday cottages:

> [W]e started making [the holiday cottages] green, because we thought it would be an attraction. . . . And we found that actually the systems that we were putting in were just that little bit too different, so that somebody coming to stay never learnt how to use it properly.
>
> (Interview, June 2002)

A transitions theory perspective: the 2016 zero carbon homes policy

It was first proposed in 2006 in a UK Government policy consultation document "Building a greener future" (DCLG 2006) that by 2016 all new homes built in England and Wales would be zero carbon. The house building industry was thereby given a decade to plan and respond to this ambitious target. The government recognised the degree of innovation and change that would be required to implement the policy, and this was the rationale behind allowing a ten year period for compliance:

> With these proposals the UK will becomc the first country to set a timetable for delivering zero carbon homes. . . . Setting a timetable now for the progressive tightening of environmental standards over the next decade will also provide certainty for business, driving innovation in the market and reducing costs of technologies.
>
> (DCLG 2006: 1)

A transitions theory perspective is useful here, because it is just this type of major shift in a sociotechnical system that the theory aims to explain (see Geels 2011; Markard *et al.* 2012; also, the introduction to Part I). Before concluding the chapter, therefore, it is briefly explored how and why, with the passage of time, the framing and reframing of sustainable housing as low carbon played a role in the development and implementation of the 2016 zero carbon homes policy (hereafter 'the 2016 policy'), for it was shortly after the period of low carbon reframing discussed above that this radical shift in policy took place.

There is evidence that misunderstandings about how early sustainable homes were built and operated gave the UK Government an overly optimistic view on the feasibility of the 2016 zero homes target (Lovell 2014). In other words, the government and other key decision-makers behind the 2016 policy did not fully recognise some of the problems (social and technical) with the zero carbon homes that already existed in the UK (Lovell 2007a, 2007b). As explored earlier, the low reframing that had been underway in the years leading up to 2006, when the 2016 policy was announced, was led by a discourse coalition: existing sustainable homes were thus used selectively to support an argument, and not as a body of experimental evidence. In other words, there was little if any objective assessment of what worked well and what did not work with these early sustainable and low energy homes (Lovell 2007a). This shaky foundation has led to problems and setbacks in the implementation of the 2016 policy: in the period since 2006 there has been significant and protracted contestation and resistance to the 2016 policy by house builders and related housing professionals, as well as the government (which has changed since 2006 from a Labour Government to a Conservative-Liberal Coalition). This resistance has taken the form of complex struggles over the precise definition of a 'zero carbon home'. Through a number of amendments to the definition in the period 2006–2014, the official government meaning of a 'zero carbon home' has significantly changed, such that for many it now no longer actually means 'zero carbon' (Nichols 2011; WWF 2011).

A pivotal decision was made in the government's March 2011 Budget that fundamentally changed the definition. Hidden in a footnote to the Budget, and expanded on in an accompanying document "Plan for growth", was a significant shift in the definition of zero carbon, effectively reducing its stringency by one-third (HM Treasury and BIS 2011). The change was that cooking and plug-in electrical appliances (ovens, computers, fridges and so on), so-called 'unregulated energy use' (i.e. not covered by the Building Regulations), were excluded from the definition. The Treasury explained its change in approach to the zero carbon home as follows:

> 2.299 The Government will introduce more realistic requirements for on-site carbon reductions, endorsing the Zero Carbon Hub's expert recommendations on the appropriate levels of on-site reductions as the starting point for future consultation, along with their advice to move to an approach based on the carbon reductions that are achieved in real life, rather than those predicted by models. This will be complemented by cost-effective options for off-site carbon reductions, relative to the Government's pricing of carbon, and Government will work with industry through consultation on how to take this forward.
>
> 2.300 This approach will deliver zero-carbon homes on a practical basis from 2016, with significantly reduced costs to industry, compared to previous proposals.
>
> (HM Treasury and BIS 2011: 117)

There are many interesting elements of this statement, not least the underlying economic theme, with the use of terms such as "cost-effective options" and delivery of zero carbon homes "on a practical basis . . . with significantly reduced costs", thereby echoing the ecologically modern thrust of the life cycle storyline discussed above.

There was a fierce reaction to the change in definition, with WWF (an NGO that played an active role in UK sustainable and zero carbon homes policy since the late 1990s) resigning from the UK Zero Carbon Home Taskforce in protest, stating in their resignation letter to the Housing Minister that: "... the alteration to this policy is a fundamental one which significantly undermines the original intent of this policy – for new homes to add zero net carbon emissions" (WWF 2011); and the UK Green Building Council Chief Executive Paul King commenting: "A zero carbon home will no longer do what it says on the tin" (cited in Nichols 2011).

Of particular interest here in the "Plan for growth" statement is the aim to "move to an approach based on the carbon reductions that are achieved in real life, rather than those predicted by models" (HM Treasury and BIS 2011: 117). This highlights the agency of actually existing low carbon technologies and materials in establishing legitimacy and credibility for those advocating action to mitigate climate change. Further, it hints at the role played by existing zero carbon homes in the development and implementation of the 2016 policy. A type of discursive treatment of existing low and zero carbon homes was a key feature of policy debates around the turn of the century (as discussed earlier), and it continued after the 2016 policy was announced. For example, the Green Building Council (GBC) – a non-profit consultancy, with membership from across the building industry – was asked by the UK Government to investigate the definition of a zero carbon home and give advice about how to proceed. Existing zero carbon homes are used in the GBC report to give credibility and legitimacy to its recommendations. The forty-page report, entitled simply "The definition of zero carbon" (GBC 2008), is notable for the care it takes to outline the methodology behind their recommendations. Even in the report's "Executive summary", there is a dedicated section on "Case studies and modelling scenarios", which explains how:

> *The task group began by undertaking a review of real case studies* where developers have built or were in the process of building exemplar low- and zero carbon developments to assess what solutions were being employed and to identify constraints.
>
> (GBC 2008: 4 [emphasis added])

Further, in the main body of the report by way of introduction it is explained how "[t]he UK-GBC uses a ground-breaking approach for establishing an evidence-base for future policy drawing together teams from within its membership ..." (2008: 9) and that "[t]he approach adopted was *to establish a clear fact base of relevant information*, drawn from a combination of *real case studies of zero carbon developments* of varying sizes and computer modelling" (2008: 10 [emphasis added]). These descriptions of the methodology behind their recommendations provide an indication of *how* the zero carbon home was defined – namely, through reference to, and using evidence from, existing zero carbon homes.

Whilst at first glance this case study looks to be a classic example of the 'mainstreaming' of ideas and technologies tested first within small-scale niches (the one-off sustainable and zero carbon homes built from the 1970s to early 2000s), which then become part of a new low carbon housing sociotechnical regime (through the implementation of the 2016 policy), the reality is more complex. It appears that the innovation niches in this case did not necessarily provide a solid evidence base for mainstreaming: the innovations trialled were not experimental, in the sense that they were objectively assessed; and there was little evidence of widespread learning subsequently taking place (Lovell 2007b, 2009). Instead, existing sustainable and zero carbon homes played a pivotal *discursive* role in the construction of a new zero carbon UK house building sector. It is a finding that also emerges from analysis of carbon offset markets (see Chapter 2), and together these two case studies problematise the objective, experimental role of innovation niches as conceptualised within transitions theory. They demonstrate that innovation niches are, in reality, highly politicised arenas, where the experimentation that takes place can be reframed in quite different ways, often by other actors not centrally involved in the innovation niche, as has been the case with UK low carbon housing.

Summary and conclusions

To summarise, this chapter first described, and endeavoured to explain, changes that took place in the sustainable housing sector in the UK during the 1990s and early 2000s. It then briefly explored, drawing on transitions theory, the development of a significant new UK climate change policy: that all new homes in England and Wales will be zero carbon by the year 2016. Existing sustainable housing built by a deep green advocacy coalition was reframed during the late 1990s and early 2000s as a solution to several policy problems. These different framings of what the policy problem was with respect to housing, and therefore its solutions, have been outlined and explored. A growing government focus on climate change mitigation at the time – including ambitious target setting – encouraged the framing of existing sustainable housing as a solution to climate change. Thus, sustainable housing was reframed as low carbon housing by the UK Government and other more mainstream organisations. The evaluation of two key storylines used by the low carbon discourse coalition – life cycles and smart housing – demonstrates how attempts to reframe the existing practices of sustainable housing as low carbon were about crafting a new way of talking and thinking about these homes, centred on notions of economic efficiency and innovation. This chapter provides evidence of the politics of low carbon framing and reframing, showing, for example, tensions around the use of low carbon storylines.

Further, the chapter also demonstrates how materials and technologies are a key component of frames. In the majority of cases, the low carbon discourse coalition were reframing sustainable housing actually built by members of the sustainable housing advocacy coalition. In other words, the low carbon discourse coalition borrowed material evidence – the sustainable housing – from the advocacy coalition. The houses were used as evidence to prove that the ideas and technologies embedded within them work, thereby giving instant credibility to what otherwise might have been dismissed as rhetoric. With the legitimacy of low and zero carbon homes significantly bolstered in this way, the government felt able to propose a radical new policy in 2006 – namely, that all new homes built in England and Wales would be zero carbon within a decade. Explanations as to why this ambitious policy has faltered have their origins in the low carbon reframing that took place at the turn of the century – in particular, the ways in which low carbon homes were used selectively as a key part of discursive storylines. This chapter shows how low carbon economies can be made from a targeted and particular *re*framing of existing practices and ideas. The process of making low carbon economies in existing areas of activity and policy (housing) is therefore about stitching together the old and the new; taking some ideas, technologies and institutions and actively discarding others.

Notes

1 Chapter 4 is based on a combination of interview and documentary evidence. Approximately fifty in-depth semi-structured interviews were conducted in the period 2001–2003 with a range of people involved in sustainable housing and climate change policy and practice in the UK. Organisations interviewed include local and national government, sustainable housing groups, consultancies, Registered Social Landlords (RSLs), non-governmental organisations, regional government agencies and private sector house builders. The interviews were transcribed and coded, in order to identify key policy discourses. In addition, documentary evidence was compiled and analysed from a range of sources, including government policy documents, housing and energy industry trade magazines and the national press.

2 I have also heard this story with a BMW car substituted for the expensive kitchen (Source: Interview with sustainable housing project manager, March 2003).

Bibliography

Barton, H. (1998). "Eco-neighbourhoods: a review of projects". *Local Environment* **3**(2): 159–177.

Bhatti, M., J. Brooke and M. Gibson (1994). "Housing and the new environmental agenda: an introduction". *Housing and the Environment: An Introduction*. M. Bhatti, J. Brooke and M. Gibson (eds). Southampton, Chartered Institute of Housing: 1–13.

Brown, P. (2003). "Tower power: why 1960s monoliths have a future". *Guardian* (16 April): 11.
Building Research Establishment Ltd (BRE) (2003). "Clear skies – renewable energy grants". Online at: www.clear-skies.org (accessed 21 May 2003).
Bulkeley, H. and M. Betsill (2003). *Cities and Climate Change: Urban Sustainability and Global Environmental Governance*. London and New York, NY, Routledge.
Centre for Alternative Technology (CAT) (1995). "Crazy idealists? The CAT story". Machynlleth, The Centre for Alternative Technology.
Chappells, H. and E. Shove (2000). "Sustainable homes and integration". *Domestic Consumption Utility Services and the Environment (DOMUS)*. H. Chappells, M. Klintman, A.-L. Linden, E. Shove, G. Spaargaren and B. van Vliet (eds). Wageningen, Universities of Lancaster, Wageningen and Lund: 105–128.
Christoff, P. (2000). "Ecological modernisation, ecological modernities". *Environmental Politics* **5**(3): 476–500.
Dawling, P. (1992). "What is the Communes Network?" Online at: www.diggersand-dreamers.org.uk (accessed 22 May 2003).
Department for Communities and Local Government (DCLG) (2006). "Building a greener future: towards zero carbon development". Online at: http://webarchive.nationalarchives.gov.uk/20120919132719/www.communities.gov.uk/documents/planningandbuilding/pdf/153125.pdf (accessed 2 May 2014).
Department for the Environment, Food and Rural Affairs (DEFRA) (2000a). "Transport 2010 – the ten year plan". London, Department for the Environment Food and Rural Affairs (DEFRA).
Department for the Environment, Food and Rural Affairs (DEFRA) (2000b). "Warm Homes and Energy Conservation Act". London, HMSO.
Department for the Environment, Food and Rural Affairs (DEFRA) (2001). "United Kingdom fuel poverty strategy". London, HMSO.
Department for Transport, Local, Government and the Regions (DTLR) (2002). "Approved document L1 – The Building Regulations 2000 (Conservation of fuel and power in dwellings)". London, HMSO.
Department of the Environment, Transport and the Regions (DETR) (2000). "Climate change: the UK programme". London, HMSO.
Department of Trade and Industry (DTI) (2002). "First stage of major PV demonstration programme launched". *New Review* **52**: 1.
Department of Trade and Industry (DTI) (2003a). "Our energy future – creating a low carbon economy: Energy White Paper". Online at: www.berr.gov.uk/files/file10719.pdf (accessed 11 December 2003).
Department of Trade and Industry (DTI) (2003b). "Press release: construction industry must build a greener future". Online at:www.gnn.gov.uk/ (ref P/2003/244) (accessed 9 April 2003).
Desai, P. (2003). "Bioregional and BedZed". Paper presented at the Sustainable Homes Renew, Refurbish, Regenerate Conference, Westminster, 6 November.
Dobson, A. (2000). *Green Political Thought*. London and New York, NY, Routledge.
Dryzek, J. (1997). *The Politics of the Earth: Environmental Discourses*. Oxford, Oxford University Press.
Ekins, P. (2002). "Sustainable buildings and the environment". London, Forum for the Future, The Housing Corporation.
English Partnerships (2003). "Millennium communities programme". Online at: www.englishpartnerships.gov.uk (accessed 21 May 2003).

Environmental Data Services (ENDS) (2001). "Private developers shun Building Research Establishment's Ecohomes scheme". *ENDS* **320**(September): 31–32.

Environmental Resources Management (ERM) (2002). "Fiscal incentives for sustainable homes". London, WWF-UK.

European Eco-village Network (2003). "What are ecovillages?" Online at: http://europe.ecovillage.org/uk/network/index.htm (accessed 23 April 2003).

Findhorn Ecovillage (2003). "Findhorn – an introduction". Online at: www.ecovillagefindhorn.com/ (accessed 23 April 2003).

Geels, F. W. (2011). "The multi-level perspective on sustainability transitions: responses to seven criticisms". *Environmental Innovation and Societal Transitions* **1**(1): 24–40.

Generating Solar Homes (2002). *Source Newsletter – Generating Solar Homes*. Nottingham, Nottingham Energy Partnership.

Grubb, M. J. (2002). "Britannia waives the rules: the United Kingdom, the European Union and climate change". *New Economy* **9**(3): 139–142.

Hajer, M. A. (1995). *The Politics of Environmental Discourse: Ecological Modernisation and the Policy Process*. Oxford, Clarendon Press.

Her Majesty's (HM) Government (2008). "Climate Change Act 2008". Online at: www.legislation.gov.uk/ukpga/2008/27/contents (accessed 7 May 2014).

Her Majesty's (HM) Treasury and Business Innovation and Science (BIS) (2011). "The plan for growth". London, HM Treasury and UK Department for Business, Innovation and Science.

Hockerton Housing Project (2003). "Hockerton media coverage". Online at: www.hockerton.demon.co.uk/media/index.html (accessed 23 April 2003).

Kingdon, J. W. (2003). *Agendas, Alternatives and Public Policies*. New York, NY, Harper Collins College Publishers.

Liddell, H. and N. Grant (2002). "Eco-minimalism: getting the priorities right". *Building for a Future* **12**: 10–13.

Lovell, H. (2004). "Framing sustainable housing as a solution to climate change". *Journal of Environmental Policy and Planning* **6**(1): 35–56.

Lovell, H. (2007a). "Exploring the role of materials in policy change: innovation in low energy housing in the UK". *Environment and Planning A* **39**(10): 2500–2517.

Lovell, H. (2007b). "The governance of innovation in socio-technical systems: the difficulties of strategic niche management in practice". *Science and Public Policy* **34**: 35–44.

Lovell, H. (2009). "The role of individuals in policy change: the case of UK low energy housing". *Environment and Planning C* **27**: 491–511.

Lovell, H. (2014). "The making of a zero carbon home". Paper presented at the International Workshop: Devices and Desires – The Cultural Politics of a Low Carbon Society. J. Stripple, H. Bulkeley and M. Paterson (eds). Lund, Sweden, 21–23 May.

Lovell, H., H. Bulkeley and S. E. Owens (2009). "Converging agendas? Energy and climate change policies in the UK". *Environment and Planning C: Government and Policy* **27**(1): 90–109.

Lowenstein, O. (2001). "From BedZed to eternity". *Building for a Future* **11**: 16–21.

Markard, J., R. Raven and B. Truffer (2012). "Sustainability transitions: an emerging field of research and its prospects". *Research Policy* **41**(6): 955–967.

Mol, A. P. J. and G. Spaargaren (2000). "Ecological modernisation theory in debate: a review". *Ecological Modernisation around the World: Perspectives and Critical Debates*. A. P. J. Mol and D. A. Sonnenfeld (eds). London and Portland, OR, Frank Cass: 17–49.

Newell, P. (2000). *Climate for Change: Non-State Actors and the Global Politics of the Greenhouse*. Cambridge, Cambridge University Press.

Nichols, W. (2011). "Budget 2011: green builders furious at watering down of 'zero-carbon' home standards". Online at: www.businessgreen.com/bg/news/2036823/budget-2011-green-builders-furious-watering-zero-carbon-home-standards (accessed 14 August 2014).

Office of the Deputy Prime Minister (ODPM) (2003a). "The decent homes target implementation plan". London, HMSO.

Office of the Deputy Prime Minister (ODPM) (2003b). "Sustainable communities: building for the future". London, HMSO.

Porter, G. and J. W. Brown (1996). *Global Environmental Politics*. Boulder, CO, Westview Press Inc.

Pragnell, M., L. Spence and R. Moore (2000). *The Market Potential for Smart Homes*. York, The Joseph Rowntree Foundation.

Royal Commission on Environmental Pollution (RCEP) (2000). "Energy – the changing climate, 22nd report". London, Royal Commission on Environmental Pollution.

Rudlin, D. (2002). "Green housing". Paper presented at the Green is the New Black Seminar, The Building Centre Trust, London, 18 November.

Sabatier, P. A. (1998). "The advocacy coalition framework: revisions and relevance for Europe". *Journal of European Public Policy* **5**(1): 98–130.

Sabatier, P. A. and H. C. Jenkins Smith (eds) (1993). *Policy Change and Learning: An Advocacy Coalition Approach*. Boulder, CO, Westview Press.

Sandbach, F. (1980). *Environment, Ideology and Policy*. Oxford, Basil Blackwell.

Smith, M., J. Whitelegg and N. Williams (1998). *Greening the Built Environment*. London, Earthscan.

Sustainable Development Commission (SDC) (2003). "Policy audit of United Kingdom climate change policies and programmes". London, Sustainable Development Commission (SDC)/The Edinburgh Centre for Carbon Management Ltd.

The Ecologist (1972). *Blueprint for Survival*. London, The Ecologist.

The Housing Forum (2001). "Homing in on excellence: a commentary on the use of offsite fabrication methods for the United Kingdom house building industry". London, The Housing Forum.

The Housing Forum (2002). "Six guiding principles to improve the sustainability of the housing construction industry". London, The Housing Forum.

The West Stevenage Consortium (2002). "West Stevenage homepage". Online at: www.weststevenage.co.uk/ (accessed 7 April 2002).

Tickle, L. (2003). "Urban clearway". *Guardian* (29 October): 12–13.

Town & Country Planning Association (TCPA) (2002). "Sustainable housing forum interim report". London, Town & Country Planning Association.

United Kingdom Green Building Council (UK-GBC) (2008). "The definition of zero carbon". Online at: www.ukgbc.org/resources/publication/uk-gbc-task-group-report-definition-zero-carbon (accessed 2 May 2014).

Vale, B. and R. Vale (1980). *The Self Sufficient House*. London, Macmillan.

Weale, A. (1992). *The New Politics of Pollution*. Manchester and New York, NY, Manchester University Press.

Wood, A. (1990). "History and overview of communal living". Online at: www.diggers-anddreamers.org.uk/Articles/199001.htm (accessed 22 April 2003).

World Wide Fund for Nature (WWF) (2003). "One million sustainable homes – turning words into action". Online at: www.wwf-uk.org/filelibrary/pdf/sustainablehomes_dec02.pdf (accessed 6 December 2004).

World Wide Fund for Nature (WWF) (2011). "Letter to the government: resignation from the zero carbon taskforce". Online at: http://assets.wwf.org.uk/downloads/zctf_resignation_letter.pdf (accessed 2 May 2014).

5

THE FRAMES OF CARBON ACCOUNTING[1]

Key findings

- Carbon accounting is a new, rapidly expanding area of practice and expertise, brought into being by the emergence of climate change as a significant problem.
- There are multiple frames of carbon accounting – physical, political, market-enabling, financial and social/environmental – comprising divergent understandings and practices.
- Carbon accounting is important in the making of low carbon economies, because it forms the calculative bedrock.
- Unresolved tensions in carbon accounting have the potential to undermine confidence in climate science, policies, markets and reporting, thereby ultimately discouraging action to mitigate climate change.
- Attention to framing helps us understand how several different understandings of carbon accounting can continue to co-exist (because, in some cases, the frames have little intersection) and highlights some of the problems of multiple frames.

Introduction

To scientists, carbon accounting is "the practice of making scientifically robust and verifiable measurements of GHG [greenhouse gas] emissions" (Watson 2009: 6). To political negotiators, it implies "the rules for comparing emissions and removals as reported with commitments" at a national level (IPCC 2005: 265). To practitioners in the United Nations Clean Development Mechanism (CDM) market (see Chapter 2), it involves the measurement of reductions in emissions relative to a hypothetical baseline and other processes associated with the subsequent creation of a new tradable commodity: a carbon credit (Ministry of the Environment Japan 2009). To the International Accounting Standards Board (IASB) (see Chapter 7), it concerns the accounting of tradable emission rights and obligations arising under Emissions Trading Schemes. To the increasing numbers of companies reporting to the Carbon Disclosure Project (CDP), The Climate Registry or other similar schemes, it involves the measurement and disclosure of greenhouse gas emissions for which companies accept varying degrees of responsibility (WBCSD and WRI 2004; PricewaterhouseCoopers 2007; Kolk *et al.* 2008; DEFRA 2009).

On a personal level, these different meanings and understandings of carbon accounting became apparent to me when in 2009 I accepted an offer to sit on the Technical Working Group of an international carbon accounting and disclosure standards organisation. At the time I was heavily engaged in research on carbon financial accounting (see Chapter 7), and I was initially confused that none of the rest of the Technical Working Group appeared to be interested in this issue. Perhaps I was just being a little too enthusiastic; maybe the fact that emission allowances did not fit well into any of the existing International Accounting Standards was rather ordinary and mundane to my expert practitioner and policy colleagues on the working group? But no, it turned out that this was a type of carbon accounting most of them genuinely had not come across before. The 'frame' of carbon accounting they were engaged in was a different one – namely, social/environmental carbon accounting – and financial carbon accounting was a another world to them altogether.

This experience was my first indication of the numerous definitions or frames of carbon accounting in use: physical, political, market-enabling, financial and social/environmental. Carbon accounting in these different senses has become an essential 'enabler' of several of society's key responses to the problem of climate change, including national emission limitation commitments, corporate climate change performance targets and carbon markets. Yet its role and contribution is generally overlooked. Furthermore, the connections, overlaps and discontinuities between different forms of carbon accounting have not received sufficient critical attention: different manifestations of carbon accounting each tend to have their own institutions, normative practices and distinctive discourses, including academic literatures.

This chapter takes a holistic view of what carbon accounting means across disciplines and institutions, in order to make sense of the differences by placing different forms of carbon accounting in their historical and social contexts. The concept of framing is useful to help understand why different conceptions of carbon accounting have developed, why certain issues are hotly contested (whereas others are not) and why carbon accounting practices frequently fall short of expectations. Interestingly, unlike sustainable and low carbon housing explored in Chapter 4, there has, to date, been little direct competition between the carbon accounting frames. This is because they originate from different spheres or communities of expertise and, until recently at least, have operated side-by-side in relative harmony, in large part because they are so different. This chapter argues that by explicitly acknowledging and highlighting the framing of carbon accounting by these distinct communities, there are significant opportunities to encourage constructive learning and policy change. At present, unrecognised and unresolved tensions in carbon accounting are undermining confidence in climate science, policies, markets and reporting, thereby ultimately discouraging action to mitigate climate change. Attention to framing helps us better understand the situation and illuminates practical solutions.

An account of the five frames forms the empirical core of the chapter. The methodology used to identify and explore the five frames is primarily a literature review of 'grey' or policy literature, as well as academic papers. It is also informed by my previous role as a member of the Technical Working Group (TWG) of an international carbon accounting organisation (2009–2012) and by Dr Francisco Ascui's experience in government, carbon markets and carbon management consultancy. The TWG membership, including attending monthly committee meetings and providing expert input to reports and strategy, has provided valuable empirical context.

The chapter shows that an extensive literature on physical carbon measurement pre-dates and influences the more technical literature on political accounting of carbon in national inventories, which in turn influences subsequent market-enabling carbon accounting. Only in recent years, as carbon markets have begun to have material impacts on company balance sheets, have financial accountants started to address the financial accounting of rights and obligations in those markets (see Chapter 7). A largely separate literature critically examines issues around corporate carbon disclosure and reporting (see Kolk *et al.* 2008), building on a wider field of literature on social and environmental accounting. At the time of analysis for this chapter (2009–2010), there was only a small (but valuable) amount of research directly assessing the politics and practices of carbon accounting from a broader perspective (Bebbington and Larrinaga-Gonzalez 2008; Cook 2009; MacKenzie 2009).[2] In general, as noted, debates over carbon accounting have been taking place within frames, with relatively few interconnections. There has thus far been an absence of overt politics between frames, because of the lack of interconnections, except in some particular instances, explored later in the chapter. Here, by bringing together and critically examining the different meanings of carbon accounting, it is hoped that this chapter demonstrates the advantages of a more holistic assessment.

The chapter is structured as follows. First, an account of the five distinctive framings of carbon accounting is provided, comprising: physical, political, market-enabling, financial and social/environmental carbon accounting. Within this, examples of key tensions in different understandings of carbon accounting are discussed that illustrate the complexity of the issues under consideration.

Second, a working definition of carbon accounting is proposed, representing the contemporary range of meanings encompassed by the term, which constitutes a useful framework within which various problematic accounting issues can be situated and clearly identified. Third, in conclusion, it is demonstrated how unresolved issues in carbon accounting have material negative consequences, and thus an improved understanding of the underlying causes of friction may contribute to finding workable solutions to climate change.

The multiple frames of carbon accounting

Over time, a range of actors and disciplines have attempted to measure 'carbon' and its impacts in various ways, for a number of different reasons. Five major framings of carbon accounting are identified, involving actors as diverse as scientists, politicians, economists, accountants and activists. It is shown how three of these – physical, political and market-enabling carbon accounting – are closely related to one another, developing in sequence and each relying on the earlier frame. The fourth, financial carbon accounting, also follows in roughly temporal sequence as a consequence of market-enabling carbon accounting, but has very different origins and objectives and is largely blind to the earlier frames. By contrast, the fifth frame of social/environmental carbon accounting has a longer pedigree, which runs alongside the other frames, sometimes interacting, but with its own specific origins and objectives. These five frames are not exclusive of other framings, and no doubt each can be critiqued from a variety of further perspectives. Nor can this chapter provide an exhaustive summary of the literature in relation to any individual frame. Rather, the aim is to provide sufficient evidence to demonstrate that such framings exist, point to some of the key institutions, actors and social contexts that make up each of the identified frames and thereby demonstrate the ways in which carbon accounting provides a foundation for the making of low carbon economies (see also Chapter 7).

Physical carbon accounting

The first of the frames can be characterised as the natural sciences view of carbon accounting as a matter of physical measurement, estimation or calculation and attribution of greenhouse gas fluxes through the biophysical environment. It has a long history: the first quantitative account of the global carbon cycle, including an estimate of the human-induced contribution from the combustion of fossil fuels, was given by the Swedish geologist Arvid Hogbom in 1895, and later used by his chemist colleague Svante Arrhenius to postulate the theory that the latter activity could cause long-term warming of the global climate in a seminal 1896 paper (Hogbom 1895; Arrhenius 1896; Rodhe *et al.* 1997). Although Arrhenius was only aware of two greenhouse gases at the time (carbon dioxide and water vapour), his estimate of the potential warming associated with a doubling of greenhouse gas concentrations in the atmosphere (5.78C) was surprisingly close to modern-day estimates (Arrhenius 1896; see IPCC 2007). However, he believed that such an outcome would not eventuate for many thousands of years, based on Hogbom's data on contemporary emission rates. As long as the implications of carbon accounting were believed to be benign or at worst remote, it remained a topic primarily of interest to geologists and atmospheric chemists seeking to understand natural processes, such as the causes of past ice ages.

However, by the 1960s, increasingly accurate instrumental measurements of atmospheric carbon dioxide levels being made at the Mauna Loa Observatory (MLO) confirmed that concentrations were higher than pre-industrial levels, and rising (Pales and Keeling 1965). By the 1980s, scientific concern about human-induced global warming had well and truly 'overflowed' the purely scientific frame to become a subject of intense political and economic debate. This debate took place at multiple levels, the most significant of which was the United Nations General Assembly, where a number of resolutions eventually led to UN General Assembly Resolution 45/212 in 1990, which initiated negotiations that concluded in the adoption of the UN Framework Convention on Climate Change (UNFCCC) at the Rio Earth Summit in 1992 (United Nations 1990). This can be seen as the founding moment for the second frame of reference, political carbon accounting.

Physical carbon accounting continues to be the primary frame of reference on carbon accounting for thousands of climate scientists worldwide (for a broad synthesis of the literature, see Chapter 2 in IPCC 2007). Nevertheless, it is increasingly difficult, if not impossible, to maintain separation between the science and the politics of climate change, as demonstrated by the furore generated over leaked emails from the University of East Anglia's Climatic Research Unit (CRU) in the lead-up to the Copenhagen climate change summit in late 2009 (Biello 2010).

With the exception of financial carbon accounting, all of the other framings look to physical carbon accounting for fundamental principles. Tensions and inconsistencies arise for two main reasons. First, non-scientists can be frustrated by the inability of science to give definitive answers in certain areas, such as the magnitude of non-carbon dioxide impacts from air travel, which has led to wide divergence in estimates of air travel offset requirements (see Padgett *et al.* 2008; DEFRA 2009). Second, the provisional, evolving nature of the science poses a challenge for other forms of carbon accounting, which seek to arrive at final conclusions with fixed consequences, several examples of which are provided in the discussion of the other framings of carbon accounting.

Political carbon accounting

The new political framing of climate change represented by the UNFCCC required a corresponding reframing of physical carbon accounting to suit an array of new objectives, including the attribution of "common but differentiated responsibilities"[3] (UNFCCC Article 3.1) – words entailing significant economic consequences. The Intergovernmental Panel on Climate Change (IPCC), which was established in 1988 as the scientific and technical advisory body to the ongoing climate negotiations, played a key role in this reframing process (Fogel 2005). The IPCC is a classic example of a boundary organisation that links and mediates between scientific and policy institutions and actors (Jasanoff *et al.* 1995; Guston 2007). In fact, it produces explicitly hybrid knowledge that is neither purely scientific nor purely political, but both: the major IPCC reports comprise both a summary of the scientific literature prepared by a committee of scientists and a summary for policymakers that is only finalised in the highly charged political arena of a UNFCCC plenary, to which all states are invited (Miller 2001; Fogel 2005).

The UNFCCC made carbon accounting at the national level mandatory for all signatories (termed 'Parties'). Article 4.1 (a) requires all Parties to

> [d]evelop, periodically update, publish and make available ... national inventories of anthropogenic emissions by sources and removals by sinks of all greenhouse gases not controlled by the Montreal Protocol, using comparable methodologies to be agreed upon by the Conference of Parties ...
>
> (United Nations 1992)

The IPCC was charged with developing the necessary "comparable methodologies" (see also Chapter 6 for discussion of the role of the IPCC in forest carbon accounting). The first IPCC Guidelines for National Greenhouse Gas Inventories were duly produced in 1995 and soon replaced by the "Revised 1996 guidelines for national greenhouse gas inventories" (IPCC 1996). Use of the "Revised 1996 guidelines" was subsequently mandated for national carbon accounting under both the UNFCCC and its subsidiary instrument, the Kyoto Protocol, in 1997. As Miller observes:

> Measures of national emissions of greenhouse gases have become the accepted means within the climate regime for assigning blame for changes in the climate and therefore for assigning responsibility for undertaking action to help stabilize the atmosphere. Such measures thus have enormously high political significance within the regime ...
>
> (2001: 489)

The role of boundary organisations such as the IPCC is to come up with both the normative and technical judgements required to produce standardised and politically acceptable carbon accounting rules, methodologies and procedures (Miller 2001). Thus, the political framing of carbon accounting takes a step away from the scientific mode of measurement, calculation and estimation of greenhouse gas emissions at the global level, towards a function of monitoring and reporting at the national level. Political expediency dictates the scope of national inventories: emissions that cannot be attributed to human activities, emissions of greenhouse gases already controlled by the Montreal Protocol and emissions associated with international air and maritime transport are all excluded (IPCC 1996). The need for standardised methodologies to enable comparisons between countries and over time creates the potential for conflict with the provisional and ever-evolving nature of the science (see again Chapter 6 for a detailed case study of this in relation to forest carbon markets).

An apt illustration of such a conflict concerns the use of conversion factors to evaluate the net impact of different greenhouse gases (GHGs), each with their own unique atmospheric chemistry and contribution to global warming (IPCC 2007; MacKenzie 2009; Plattner *et al.* 2009). Climate scientists have formulated various ways of measuring and commensurating the climate impacts of different GHGs, the most influential of these being Lashof and Ahuja (1990), who developed the theoretical framework for what is now known as 'global warming potential' (GWP) – a metric of the contribution to global warming of a given mass of GHG over a given time horizon, all conveniently expressed in multiples of carbon dioxide equivalent. The driver for development of this index was both political and economic, as Lashof and Ahuja (1990: 529) note: "An index to compare the contribution of various 'greenhouse' gas emissions to global warming is needed to develop cost-effective strategies for limiting this warming". In addition to being only one of several possible approaches (for others, see Plattner *et al.* 2009), the approach is beset with uncertainties, both empirical (e.g. uncertainty in observations of atmospheric longevity) and conceptual (e.g. results being sensitive to the choice of time horizon). However, in 1997, Article 5.3 of the Kyoto Protocol mandated the use of an arbitrary set of global warming potentials (those published in 1996 by the IPCC in its Second Assessment Report) for the purposes of national carbon accounting over the first commitment period (2008–2012). This political decision has given rise to divergences between physical and political carbon accounting. Estimates of the GWP of various GHGs published in the scientific literature, and summarised by the IPCC in subsequent assessment reports, continue to be revised, whereas the factors now used in reporting under the UNFCCC and Kyoto Protocol – and in a wide range of national and corporate reporting standards developed since then – have remained static (UNFCCC 2006, 2008). The UK's national emissions, for example, are calculated using the 1996 GWP "exchange rate" for methane of twenty-one times the equivalent mass of carbon dioxide, whereas the 2007 IPCC assessment report suggests that a value of twenty-five times is more accurate – a variation of nearly 20 per cent (DEFRA 2006; IPCC 2007: 212). Further, if measured over a twenty-year time horizon rather than the conventional 100 years, the latest GWP of methane rises to seventy-two (IPCC 2007). These alternatives have major implications for where governments should direct their climate mitigation efforts, yet the political decisions are fundamentally arbitrary from a scientific perspective. As Milne *et al.* (2010: 27) observe, after reviewing the wild fluctuations in estimates of New Zealand's national inventory from 2005 to 2009: "GHG emission accounting, like much other accounting, is set to remain part science, part modelling, part guesswork and part negotiation".

Market-enabling carbon accounting

In the 1990s, relying in part on the use of global warming potentials to enable the commensuration of different GHGs emitted in different places at different times, economists such as Nordhaus (1991: 924) began to consolidate a then already widespread economic framing of climate change (in keeping with earlier market-based attempts to mitigate environmental problems, such as the US SO_2 market; see MacKenzie 2007) by identifying climate change as an optimal control problem, the ideal policy solution to which would lie at the point where marginal abatement costs would equal the marginal damages caused by climate change. From here, it was a small – yet momentous – step to postulate that a market for abatement of greenhouse gases would be more likely to arrive at this optimal solution than even the most well-meant policymaking. The US had been experimenting with a market approach to regulating sulphur dioxide emissions since the early 1990s, with great apparent success, in terms of breaking the policymaking impasse, reducing emissions at lower than expected cost and fostering innovation (Wambsganss and Sanford 1996; Johnston *et al.* 2008; MacKenzie 2009). Largely at the insistence of the US, the individual caps on developed countries' GHG emissions in the Kyoto Protocol were linked by the three "flexibility mechanisms" of International Emissions Trading, Joint Implementation and the Clean Development Mechanism (CDM), together creating a framework for a global market in greenhouse gas emission rights, driven by emission obligations (United Nations 1998).

Discrepancies immediately arose between the political carbon accounting of the UNFCCC and the market-enabling carbon accounting of the Kyoto Protocol. Creating the demand and supply necessary for a market in something as intangible as GHG emission rights and obligations implies numerous acts of quantification, measurement and commensuration (see Part III of this book; Espeland and

Stevens 1998; Lohmann 2005, 2009; MacKenzie 2009). On the one hand, demand was created by placing caps on national emissions from developed countries, which naturally looked to existing IPCC methodologies developed for the purposes of measuring national emissions in a consistent manner, as discussed. On the other hand, supply was created in two different ways: first, by creating emission rights and enabling trading between capped countries facing different costs of compliance; second, by creating an entirely new, fictitious commodity in the form of an emission right based on an emission reduction achieved in a country without a cap (this being the function of the CDM; see Chapter 2).

Under the UNFCCC, developing countries have an obligation to account for their national emissions, but without any associated or implied responsibility (under the principle of "common but differentiated responsibilities"). With the introduction of the Kyoto Protocol's CDM, entirely new carbon accounting rules were required to enable the measurement of emission reductions against a hypothetical baseline within defined projects, whereas previous accounting rules concerned the measurement of emissions and removals taking place within national boundaries. Such emission reductions give rise to credits known as Certified Emission Reductions (CERs); a developed country may obtain such CERs and use them to exceed its cap by one tonne of carbon dioxide (or its equivalent in other GHGs) per CER.

Thus, the CDM is engaged in an entirely novel project of "making things the same" (MacKenzie 2009): in this case, making reductions in emissions against a baseline equivalent to emission rights in developed countries. The Kyoto Protocol created a mandate for this, but did not specify how it would work; more detailed rules were not agreed on at the political level until 2001 and the practical framework continues to evolve, with the full 'rulebook' now running to over 1,000 pages (CDM Rulebook 2014). Methodologies for measuring emission reductions against a hypothetical baseline simply did not exist and had to be invented – significantly, in this case, not by scientists or politicians, but by a range of non-state, largely private sector actors involved in CDM project development, via a bottom-up process of methodology proposal, review and rejection or acceptance by the CDM Executive Board. As discussed in Chapter 2, there are now literally hundreds of CDM methodologies available for different types of projects. The process has been criticised for failing to make the necessary political decisions to resolve contentious issues and for producing outcomes riddled with inconsistencies (Michaelowa *et al.* 2007).

One of the most contentious areas of carbon accounting over the past two decades has been the treatment of stored carbon, known in UNFCCC parlance as 'sinks'. Examples of sinks include carbon stored in forests (Watson 2009; see also Chapter 6), forest products (Lim *et al.* 1999), soils (Shackley and Sohi 2010) or deep underground – for example, through carbon capture and geological storage (CCS) (IPCC 2005; Grönkvist *et al.* 2006). Negotiators of the Kyoto Protocol in

1997 were unable to decide whether to allow the CDM to provide carbon credits to projects that reduce deforestation, thereby maintaining forest carbon sinks that would otherwise be lost. The IPCC was commissioned to prepare a Special Report on Land Use, Land-Use Change and Forestry (LULUCF), which highlighted the many technical difficulties associated with measuring reductions in deforestation, although not without considerable dispute between participants: one observer relates the stories of numerous "boundary battles" taking place within the IPCC Special Report plenary over the issue (Fogel 2005: 200).

When more detailed rules on the CDM were finally agreed to in Marrakesh in 2001, eligible activities in the LULUCF sector were limited to afforestation and reforestation, excluding reduced deforestation. Reduced deforestation is still excluded from carbon markets under the Kyoto Protocol, although it is on the agenda for a post-2012 climate agreement, under the new guise of Reducing Emissions from Deforestation and Forest Degradation (REDD) (see Chapter 6). As Chapter 6 outlines, reduced deforestation remains a hotly contested area in market-enabling carbon accounting; one where the collisions between the scientific, political and market-enabling frames have not yet been resolved (see also Eliasch 2008; Neeff and Ascui 2009).

Financial carbon accounting

The Kyoto Protocol created new GHG emission rights and obligations on states, not corporations. In many jurisdictions, however, states have created mirroring rights and obligations on corporations, particularly the owners or operators of large point sources of emissions such as power stations and industrial facilities, through the implementation of national or regional Emissions Trading Schemes. The most notable of these has been the European Union Emissions Trading Scheme (EU ETS), to date still the largest carbon market in the world, with transaction volumes reaching US$171 billion in 2011 (Kossoy and Guigon 2012). Companies operating in these carbon markets have new liabilities, assets and financial flows to account for in their financial reports. However, doing so has proven difficult, due to conflicts which can be characterised as the collision between a new attempt at framing carbon in terms of existing financial accounting concepts and the incumbent framing in already existing carbon markets.

Chapter 7 provides a detailed account and exploration of financial carbon accounting, using the concepts of framing, metrology, standardisation and calculation. In brief, despite initial attempts by the IASB to establish financial carbon accounting guidelines when the EU ETS first came into being in 2005, there has been no international agreement, and a diversity of accounting practices have subsequently emerged (PricewaterhouseCoopers and International Emissions Trading Association 2007; McGready 2008; Cook 2009; MacKenzie 2009). Progress towards a global standard for financial carbon accounting has been slow for a number of reasons: carbon sits between and challenges existing financial accounting standards, including IAS 20 (government grants), IAS 38 (intangible assets) and IAS 39 (financial instruments). Scholars have, moreover, interpreted the continuing ambiguity in accounting rules as illustrative of a more fundamental lack of consensus on the accounting treatment of carbon (Cook 2009; MacKenzie 2009). Carbon has been difficult to classify in part because accountants and accounting standard-setters lack a full appreciation of the production process of carbon credits: the science, politics and market-enabling rules involved in turning greenhouse gas emissions, and emission reductions, into tradable commodities. A lack of knowledge and experience can be expected to reduce over time, but a more fundamental challenge is the way in which types of knowledge and information are framed by accountants as relevant to their decision-making. As Chapter 7 outlines, accountants typically seek to understand carbon by comparison with existing, more familiar, accounting entities such as taxes, leases, subsidies and commodities, without appreciating the complexities caused by changes in climate policy or regulation, such as the shift to increased auctioning of carbon allowances from 2013 in the EU ETS, which, for example, did not receive coverage in technical IASB discussions, despite its relevance (Lovell *et al.* 2013).

Social/environmental carbon accounting

The last of the five frames emerges from the broader context of social and environmental accounting, which has developed as a rich and diverse arena for practice and research over the past four decades (Gray *et al.* 1993; Mathews 1997; Gray 2002; Parker 2005; Unerman *et al.* 2007; Owen 2008). It is clear from these reviews that social and environmental accounting, like carbon accounting, means different things to different people. Indeed, there are close parallels between what is observed in carbon accounting today and an early description by Mathews of social accounting:

> . . . the extension of social accounting measurements and disclosures is affected by confusion, measurement problems and disagreements about the legitimacy of accounting activity in this field. The confusion arises because the term social accounting is used in different ways by different groups of people and the measurement difficulties are always present in any new area; indeed, they are what accounting is all about.
>
> (1984: 200)

While there are many aspects of social/environmental reporting practice that have contributed to contemporary carbon accounting, Gray *et al.* (2007: 17) note "the almost complete absence" of carbon accounting in the social accounting literature (however, important contributions have been made since then; see Ascui 2014 for a review). The focus here is on two traditions of particular interest: corporate sustainability reporting and product life cycle analysis (LCA). Corporate sustainability reporting has long been the most prominent area of practice and research in social and environmental accounting, termed "Social Responsibility Accounting" in Mathews' (1984) early classification of the field and defined then as: "Voluntary disclosure of information, both qualitative and quantitative, made by organisations to inform or influence a range of audiences" (Mathews 1984: 204).

Corporate sustainability reporting can be seen as an extension of traditional financial reporting to include social and environmental policies and impacts, influenced since the late 1990s by the notion of the 'triple bottom line' (Milne and Gray 2007; Milne *et al.* 2009). Energy use and greenhouse gas emission statistics appeared in some of the earliest of these sustainability reports and are now routinely included as core environmental indicators under the Global Reporting Initiative (GRI). However, they constitute only a handful of the dozens of GRI core indicators: in this tradition, climate change is only one amongst many social and environmental impacts. Climate change-specific corporate reporting can be seen as an extension of this. A significant enabler was the development of the "Greenhouse gas protocol: corporate accounting and reporting standard" by the World Resources Institute and the World Business Council for Sustainable Development (WBCSD and WRI 2004). While based in part on IPCC guidelines (i.e. political carbon accounting), the GHG Protocol introduces entirely new concepts relevant only to corporate emissions, such as a division between three scopes of direct (Scope 1), electricity- and heat-related indirect (Scope 2) and other indirect (Scope 3) emissions (WBCSD and WRI 2004). Since first publication in 2001, the GHG Protocol has been incorporated into dozens of voluntary and governmental reporting guidelines, including the GRI and an international standard (ISO14064–1). However, as noted by Kolk *et al.* (2008: 738), the appearance this gives of standardisation is misleading: many of the derived guidelines modify or supplement the GHG Protocol in unique ways (see, for example, DEFRA 2009), leading to inconsistencies in global corporate reporting.

A second major development in climate change-specific corporate reporting was the establishment of the Carbon Disclosure Project (CDP) in 2000. In 2002, backed by a group of thirty-five signatory investors with US$4.5 trillion in assets, the CDP issued a call to FT500 Global Index companies for information relating to their impacts on and from climate change (Innovest 2003). By 2009, the CDP was proudly acting "[o]n behalf of 722 investors with assets of US$87 trillion" (PricewaterhouseCoopers 2013). Carbon disclosure appears to be going mainstream even faster than its corporate sustainability parent: while the number of companies registering GRI reports reached 1,000 for the first time in 2008, the same milestone was reached in terms of companies responding to the CDP in 2007, and by 2009, the number of CDP reports (2,456) was nearly double the GRI level.[4] The comparison may be a little unfair, because a CDP "response" is not necessarily complete, nor necessarily made public; while the number of companies producing reports based on GRI guidance is undoubtedly much higher than the number registering these reports with GRI. Nevertheless, the growth in carbon disclosure, particularly since 2006 through the CDP, has been astounding. The resulting data provides a rich basis for research into the relationships between disclosure, management strategies and various measures of performance, even if a preliminary analysis by Kolk *et al.* (2008: 741) suggests that so far: ". . . in spite of increasing response rates and expanding volume of the answers, there is no real evidence that the information is helpful and is being used by investors in their decision-making processes".

It also provides a basis for emergent forms of carbon benchmarking (Mackenzie *et al.* 2009; Czyz *et al.* 2010). Kolk *et al.* (2008: 722) note a number of factors that have played a part in the rapid institutionalisation of carbon disclosure, including "the convergence of business, governments, NGOs and key academic and professional constituencies around a somewhat fragmented, decentralized and market-oriented mode of carbon governance" – namely, carbon trading. This convergence on carbon markets as a dominant paradigm has undoubtedly influenced corporate behaviour in different ways. In some constituencies, carbon accounting and reporting has been imposed on companies – for example, under the EU ETS, where annual reporting of verified carbon dioxide emissions became mandatory for large emitters from 2005 (see Chapter 7). In other constituencies, carbon accounting and voluntary reporting may be driven by anticipation of future carbon regulation. Carbon accounting has also entered the discourse of strategy: measuring an organisation's carbon footprint has become a widely accepted first step in developing a corporate climate change strategy (Hoffman 2007; Lash and Wellington 2007; PricewaterhouseCoopers 2007).

The above discussion has focused on organisations, particularly private sector corporations, as the main subjects of social/environmental carbon accounting. However, the notion of the carbon footprint descends, in part at least, from the earlier, much broader, concept of the "ecological footprint" as a way of measuring and comparing the totality of environmental impacts (Rees 1992; Wackernagel and Rees 1996). The related concept of a product carbon footprint owes a great deal to the theory and methods of life cycle analysis (LCA) (for a recent review, see Finnveden *et al.* 2009). LCA has traditionally been dominated by scientists, rather than accountants. Perhaps for this reason, a recent standard (heavily derived from LCA practices) known as PAS 2050 for carbon accounting of goods and services (BSI 2008) specified the use of the latest IPCC figures for global warming potentials, thus giving rise to discrepancies between, say, a company's emissions as reported under the EU Emissions Trading Scheme or the WRI/WBCSD GHG Protocol (which both follow the Kyoto Protocol approach of using 1996 IPCC values) and the emissions associated with the company's products. Like previously mentioned national and project-level carbon accounting, LCA and corporate reporting standards can be seen as attempts to define different boundaries and responsibilities for GHG emissions, with overlapping and contested results.

Defining carbon accounting

It is no doubt evident from the analysis in this chapter that providing a neat, concise definition of carbon accounting is somewhat elusive and problematic; this, in fact, is the point of the chapter, and it echoes the struggles over defining a 'zero carbon home' in Chapter 4. It is useful to understand carbon accounting through the analogy of a jumbled landscape created by the collisions within and between multiple frames, rather than as a neatly delineated, essentialised object of inquiry (Callon 2009; Lohmann 2009). Nevertheless, it is also helpful to have a summary of the spectrum of activities that carbon accounting can involve across the different frames of reference. Accordingly, it is proposed that carbon accounting can be understood as shown in Table 5.1. By selecting and combining different terms within this table, a multitude of more specific interpretations of carbon accounting may be derived. Thus, for example, climate scientists are chiefly concerned with the estimation or direct measurement of greenhouse gas emissions and removals at the global level for research purposes. The Climate Disclosure Standards Board, on the other hand, is interested in the reporting of greenhouse gas emissions and impacts from climate change at the corporate level for voluntary disclosure purposes. Specific definitions, then, are like pathways through the landscape created by the collisions between different frames.

While Callon and others concentrate on the ways in which discrete individual frames deconstruct themselves via "overflows" (Callon 1998; Lohmann 2009; see the introduction to Part II), it is suggested that there is another level of interaction also taking place here, characterised as jostling or collisions between different frames (Lovell and Smith 2010). In recognising the interactions between separate frames, the analysis builds upon Rein and Schon's (1993) distinction between policy disagreements and policy controversies: disagreements occur within the same frame, controversies are between different frames and thus are much more difficult to resolve. By explicitly acknowledging and highlighting the framing of carbon accounting by different communities of practice, there are significant opportunities to encourage constructive learning and policy change

Table 5.1 A working definition of carbon accounting

estimation calculation measurement monitoring reporting validation verification auditing	of	carbon carbon dioxide greenhouse gas	emissions to the atmosphere removals from the atmosphere emission rights emission obligations emission reductions legal or financial instruments linked to the above trades/transactions of any of the above impacts on climate change impacts from climate change	at	global national sub-national regional civic organisational corporate project installation event product supply chain	level, for	mandatory voluntary	research compliance reporting disclosure benchmarking auditing information marketing or other	purposes

(what Rein and Schon (1993) term "frame-reflective discourse" (150); see Etzion and Ferraro (2010) for an example).

Summary and conclusions

In summary, the diverse empirical material in this chapter has drawn attention to the multiple frames of carbon accounting. In general terms, the multiple frames of carbon accounting have, to date, mostly co-existed without overt disagreements. Whilst it has been shown that carbon accounting is contested, meaning many different things to different people, these different framings have in most cases been so distinct as to not intrude upon each other. Thus far, the politics of framing in the case of carbon accounting has been muted. This may well change in the future as inconsistencies between frames, such as the ones identified in this chapter, are more clearly recognised and there are increasing attempts to resolve them. However, it is also possible that there are strategic, political reasons for the actors working in carbon accounting to keep the multiple definitions in play. It has been observed by scholars working on other environmental topics and beyond that an element of ambiguity in the meaning of terms can be helpful, as Owens and Cowell (2002: 16) observe in relation to sustainable development: "That some strange bedfellows were able to endorse sustainable development at all demonstrates the versatility of the discourse and the ability of different groups to construct it in their own image".

This chapter has shown how carbon accounting is fundamental to considering the ways in which low carbon economies are made: it is precisely the easily overlooked systems of classification, measurement, commensuration and communication that underpin society's key responses to the "super wicked problem" (Lazarus 2009) of climate change. When carbon accounting fails to provide adequately comparable information on corporate emissions, impacts and responses to enable investors to make appropriate decisions (Kolk *et al.* 2008), fails to incentivise tropical countries to reduce deforestation or prevents investment in biomass carbon capture and storage because it fails to recognise and reward negative emissions (Grönkvist *et al.* 2006), the making of new low carbon economies is threatened. Recognising and understanding the frames of carbon accounting and bringing knowledge and experience from different communities together provides significant potential for constructive learning and policy change.

Notes

1 Chapter 5 is based, with minor variations, on a paper published previously with Dr Francisco Ascui, Lecturer at the Business School, University of Edinburgh, who was the lead author of the paper, see Ascui, F. and H. Lovell (2011). "As frames collide: making sense of carbon accounting." *Accounting, Auditing and Accountability Journal.*

24(8): 978–999. It draws on his considerable experience in the period 1999–2011 working on carbon markets in the public and private sectors.

2 For a more recent review of a rapidly growing area of scholarship, see Ascui (2014).

3 The phrase 'common but differentiated responsibility' is used in connection with global problems that are of common concern including climate change, but towards which nation states have different responsibilities and capacities to act. With regard to climate change, there are two considerations, both embodied within the UNFCCC and the Kyoto Protocol: the cumulative responsibility of countries for the problem (historical as well as current responsibility); and the ability of countries to deal with the problem (in technical, social and economic terms), see http://climate.diplomacy.edu/page/cbdr-principle for further discussion.

4 See Global Reporting Initiative (2010) and Carbon Disclosure Project (2009).

Bibliography

Arrhenius, S. (1896). "On the influence of carbonic acid in the air upon the temperature of the ground". *The London, Edinburgh and Dublin Philosophical Magazine and Journal of Science as Culture* **41**: 237–276.

Ascui, F. (2014). "A review of carbon accounting in the social and environmental accounting literature: what can it contribute to the debate?" *Social and Environmental Accountability Journal* **34**(1): 6–28.

Ascui, F. and H. Lovell (2011). "As frames collide: making sense of carbon accounting". *Accounting, Auditing and Accountability Journal* **24**(8): 978–999.

Bebbington, J. and C. Larrinaga-Gonzalez (2008). "Carbon trading: accounting and reporting issues". *European Accounting Review* **17**(4): 697–717.

Biello, D. (2010). "Negating climategate". *Scientific American* **302**(2): 16.

British Standards Institute (BSI) (2008). "PAS 2050:2008. Specification for the assessment of the life cycle greenhouse gas emissions of goods and services". London, British Standards Institute (BSI).

Callon, M. (1998). "An essay on framing and overflowing: economic externalities revisted by sociology". *The Laws of the Markets*. M. Callon (ed.). Oxford, Blackwell: 244–269.

Callon, M. (2009). "Civilizing markets: carbon trading between *in vitro* and *in vivo* experiments". *Accounting, Organizations and Society* **34**(3–4): 535–548.

Carbon Disclosure Project (2009). "Global 500 report". Online at: www.cdp.net/CDPResults/CDP_2009_Global_500_Report_with_Industry_Snapshots.pdf (accessed 27 August 2014).

Clean Development Mechanism (CDM) Rulebook (2014). "Home page – welcome to the CDM Rulebook". Online at: www.cdmrulebook.org/ (accessed 3 June 2014).

Cook, A. (2009). "Emission rights: from costless activity to market operations". *Accounting, Organizations and Society* **34**(3–4): 456–468.

Czyz, K., I. Gozdowska, C. Mackenzie and S. McMahon (2010). "FTSE CDP carbon strategy index series: 350 index results report 2010". London, ENDS Carbon.

Department for Environment, Food and Rural Affairs (DEFRA) (2006). "The United Kingdom's initial report under the Kyoto Protocol". London, Department for Environment, Food and Rural Affairs (DEFRA).

Department for Environment, Food and Rural Affairs (DEFRA) (2009). "2009 guidelines to Defra/DECC's GHG conversion factors for company reporting: methodology paper for emission factors". Online at: www.defra.gov.uk/environment/business/reporting/

pdf/091013-guidelines-ghg-conversion-factors-method-paper.pdf (accessed 10 January 2010).

Eliasch, J. (2008). *Climate Change: Financing Global Forests: The Eliasch Review*. London, Earthscan.

Espeland, W. N. and M. L. Stevens (1998). "Commensuration as a social process". *Annual Review of Sociology* **24**: 313–343.

Etzion, D. and F. Ferraro (2010). "The role of analogy in the institutionalization of sustainability reporting". *Organization Science* **21**(5): 1092–1107.

Finnveden, G., M. Z. Hauschild, T. Ekvall, J. Guinee, R. Heijungs, S. Hellweg, A. Koehler, D. Pennington and S. Suh (2009). "Recent developments in life cycle assessment". *Journal of Environmental Management* **91**(1): 1–21.

Fogel, C. (2005). "Biotic carbon sequestration and the Kyoto Protocol: the construction of global knowledge by the Intergovernmental Panel on Climate Change". *International Environmental Agreements: Politics, Law and Economics* **5**(2): 191–210.

Global Reporting Initiative (GRI) (2010). "Year in review 2009/10". Online at: www.globalreporting.org/resourcelibrary/GRI-Year-in-Review-2009-2010.pdf (accessed 27 August 2014).

Gray, R. (2002). "The social accounting project: privileging engagement, imaginings, new accountings and pragmatism over critique?" *Accounting, Organizations and Society* **27**(7): 687–708.

Gray, R., J. Bebbington and D. Walters (1993). *Accounting for the Environment*. London, Paul Chapman in association with ACCA, the Chartered Association of Certified Accountants.

Gray, R., J. Dillard and C. Spence (2007). "(Social) accounting (research) as if the world matters: postalgia and a new absurdism". Paper presented at the Fifth Asia Pacific Interdisciplinary Research in Accounting Conference, Auckland, 8–10 July.

Grönkvist, S., K. Möllersten and K. Pingoud (2006). "Equal opportunity for biomass in greenhouse gas accounting of CO_2 capture and storage: a step towards more cost-effective climate change mitigation regimes". *Mitigation and Adaptation Strategies for Global Change* **11**(5–6): 1083–1096.

Guston, D. H. (2007). *Between Politics and Science: Assuring the Integrity and Productivity of Reseach*. Cambridge, Cambridge University Press.

Hoffman, A. J. (2007). *Carbon Strategies: How Leading Companies are Reducing their Climate Change Footprint*. Ann Arbor, MI, University of Michigan Press.

Hogbom, A. (1895). "Om sannolikheten for sekulara forindringar i atmosfarens kolsyrehalt". *Svensk kemisk tidskr* **5**: 169–176.

Innovest (2003). "Carbon Disclosure Project: carbon finance and the global equity markets". London, Carbon Disclosure Project.

Intergovernmental Panel on Climate Change (IPCC) (1996). "Revised 1996 IPCC guidelines for national greenhouse gas inventories". Online at: www.ipcc-nggip.iges.or.jp/public/gl/invs1.html (accessed 10 April 2008).

Intergovernmental Panel on Climate Change (IPCC) (2005). "IPCC special report on carbon dioxide capture and storage". Online at: www.ipcc.ch/pdf/special-reports/srccs/srccs_wholereport.pdf (accessed 10 April 2008).

Intergovernmental Panel on Climate Change (IPCC) (2007). "Climate change 2007: the physical science basis. Contribution of Working Group I to the fourth assessment report of the Intergovernmental Panel on Climate Change". Cambridge and New York, NY, Cambridge University Press.

Jasanoff, S., J. C. Petersen, T. Pinch and G. E. Markle (eds) (1995). *Handbook of Science and Technology Studies*. London, Sage.

Johnston, D. M., S. E. Sefcik and N. S. Soderstrom (2008). "The value relevance of greenhouse gas emissions allowances: an exploratory study in the related United States SO_2 market". *European Accounting Review* **17**(4): 747–764.

Kolk, A., D. Levy and J. Pinske (2008). "Corporate responses in an emerging climate regime: the institutionalization of carbon disclosure". *European Accounting Review* **17**(4): 719–745.

Kossoy, A. and P. Guigon (2012). *State and Trends of the Carbon Market 2012*. Washington, DC, The World Bank.

Lash, J. and F. Wellington (2007). "Competitive advantage on a warming planet". *Harvard Business Review* (March): 1–11.

Lashof, D. A. and D. R. Ahuja (1990). "Relative contributions of greenhouse gas emissions to global warming". *Nature* **344**(5): 529–531.

Lazarus, R. J. (2009). "Super wicked problems and climate change: restraining the present to liberate the future". *Cornell Law Review* **94**(5): 1153–1233.

Lim, B., S. Brown and B. Schlamadinger (1999). "Carbon accounting for forest harvesting and wood products: review and evaluation of different approaches". *Environmental Science & Policy* **2**(2): 207–216.

Lohmann, L. (2005). "Marketing and making carbon dumps: commodification, calculation and counterfactuals in climate change mitigation". *Science as Culture* **14**(3): 203–235.

Lohmann, L. (2009). "Toward a different debate in environmental accounting: the cases of carbon and cost-benefit". *Accounting, Organizations and Society* **34**(3): 499–534.

Lovell, H. and S. J. Smith (2010). "Agencement in housing markets: the case of the UK construction industry". *Geoforum* **41**(3): 457–468.

Lovell, H., J. Bebbington, C. Larrinaga and T. Sales de Aguiar (2013). "Putting carbon markets into practice: a case study of financial accounting in Europe". *Environment and Planning C* **31**: 741–757.

Mackenzie, C., D. Hikisch and S. Ivory (2009). "UK supermarkets 2009 carbon benchmark report". London, ENDS Carbon.

MacKenzie, D. (2007). "The political economy of carbon trading: a ratchet". *London Review of Books* **29**(7). Online at: www.lrb.co.uk/v29/n07/mack01_.html (accessed 10 April 2007).

MacKenzie, D. (2009). "Making things the same: gases, emission rights and the politics of carbon markets". *Accounting, Organizations and Society* **34**(3–4): 440–455.

Mathews, M. R. (1984). "A suggested classification for social accounting research". *Journal of Accounting and Public Policy* **3**(3): 199–221.

Mathews, M. R. (1997). "Twenty-five years of social and environmental accounting research: is there a silver jubilee to celebrate?" *Accounting, Auditing & Accountability Journal* **10**(4): 481–531.

McGready, M. (2008). "Accounting for carbon". *Accountancy* **July 2008**: 84–85.

Michaelowa, A., F. Gagnon-Lebrun, D. Hayashi, L. S. Flores, P. Crete and M. Krey (2007). "Understanding CDM and methodologies: a guidebook to CDM rules and procedures". London, Department for Environment, Food and Rural Affairs (DEFRA).

Miller, C. (2001). "Hybrid management: boundary organizations, science policy, and environmental governance in the climate regime". *Science, Technology & Human Values* **26**(4): 478–500.

Milne, M. and R. Gray (2007). "Future prospects for corporate sustainability reporting". *Sustainability Accounting and Accountability*, J. Unerman, J. Bebbington and B. O'Dwyer (eds). London, Routledge: 184–207.

Milne, M., A. Ball and I. Mason (2010). "The Kyoto seesaw: accounting for GHG emissions 2008 to 2012". *Chartered Accountants Journal* (February): 25–27.

Milne, M. J., H. Tregidga and S. Walton (2009). "Words not actions! The ideological role of sustainable development reporting". *Accounting, Auditing & Accountability Journal* **22**(8): 1211–1257.

Ministry of the Environment Japan (2009). "CDM/JI manual for project developers and policy makers". Tokyo, Ministry of the Environment.

Neeff, T. and F. Ascui (2009). "Lessons from carbon markets for designing an effective REDD architecture". *Climate Policy* **9**(3): 306–315.

Nordhaus, W. D. (1991). "To slow or not to slow: the economics of the greenhouse effect". *The Economic Journal* **101**: 920–937.

Owen, D. (2008). "Chronicles of wasted time?: A personal reflection on the current state of, and future prospects for, social and environmental accounting research". *Accounting, Auditing & Accountability Journal* **21**(2): 240–267.

Owens, S. E. and R. Cowell (2002). *Land and Limits: Interpreting Sustainability in the Planning Process*. London, Routledge.

Padgett, J. P., A. C. Steinemann, J. H. Clarke and M. P. Vandenbergh (2008). "A comparison of carbon calculators". *Environmental Impact Assessment Review* **28**(2): 106–115.

Pales, J. C. and C. D. Keeling (1965). "The concentration of atmospheric carbon dioxide in Hawaii". *Journal of Geophysical Research* **70**(24): 6053–6076.

Parker, L. D. (2005). "Social and environmental accountability research: a view from the commentary box". *Accounting, Auditing & Accountability Journal* **18**(6): 842–860.

Penman, J., M. Gytarsky, T. Hiraishi, T. Krug, D. Kruger, R. Pipatti, L. Buendia, K. Miwa, T. Ngara, K. Tanabe and F. Wagner (2003). "IPCC Good Practice Guidance for land use, land-use change and forestry". Institute for Global Environmental Strategies, Kanagawa, National Greenhouse Gas Inventories Programme.

Plattner, G., T. Stocker, P. Midgley and M. Tignor (2009). "IPCC expert meeting on the science of alternative metrics, The Grand Hotel, Oslo, Norway, 18–20 March 2009: meeting report". Online at: www.ipcc.ch/pdf/supporting-material/expert-meeting-metrics-oslo.pdf (accessed 29 January 2010).

PricewaterhouseCoopers (2007). "Carbon value: robust carbon management – a framework to protect and enhance shareholder value in response to climate change". Sydney, PricewaterhouseCoopers.

PricewaterhouseCoopers (2013). "Sector insights: what is driving climate change action in the world's largest companies? Global 500 climate change report 2013". Online at: www.cdp.net/CDPResults/CDP-Global-500-Climate-Change-Report-2013.pdf (15 June 2014).

PricewaterhouseCoopers and International Emissions Trading Association (2007). "Trouble-entry accounting – revisited". London, PriceWaterhouse Coopers (Pwc) and the International Emissions Trading Association (IETA).

Rees, W. E. (1992). "Ecological footprints and appropriated carrying capacity: what urban economics leaves out". *Environment and Urbanization* **4**(2): 121–130.

Rein, M. and D. Schon (1993). "Reframing policy discourse". *The Argumentative Turn in Policy Analysis and Planning*. F. Fischer and J. Forester (eds). London, UCL Press Ltd.: 145–166.

Rodhe, H., R. Charlson and E. Crawford (1997). "Svante Arrhenius and the greenhouse effect". *Ambio* **26**(1): 2–5.

Shackley, S. and S. Sohi (eds) (2010). "An assessment of the benefits and issues associated with the application of biochar to soil". Edinburgh, A Report Commissioned by the United Kingdom Department for Environment, Food and Rural Affairs, and Department of Energy and Climate Change, UK Biochar Research Centre.

Unerman, J., J. Bebbington and B. O'Dwyer (eds) (2007). *Sustainability Accounting and Accountability*. London, Routledge.

United Nations Framework Convention on Climate Change (UNFCCC) (2006). "Updated UNFCCC reporting guidelines on annual inventories following incorporation of the provisions of Decision 14/CP.11. Note by the Secretariat, decision FCCC/SBSTA/2006/9". Online at: http://unfccc.int/documentation/documents/advanced_search/items/3594.php?rec¼j&priref¼600003988#beg (accessed 18 September 2009).

United Nations Framework Convention on Climate Change (UNFCCC) (2008). "Kyoto Protocol reference manual on accounting of emissions and assigned amount". Bonn, United Nations Framework Convention on Climate Change (UNFCCC).

United Nations (1990). "United Nations General Assembly Resolution 45/212. Protection of global climate for present and future generations of mankind". Online at: www.un.org/documents/ga/res/45/a45r212.htm (accessed 10 September 2009).

United Nations (1992). "United Nations Framework Convention on Climate Change". Online at: http://unfccc.int/resource/docs/convkp/conveng.pdf (accessed 10 January 2007).

United Nations (1998). "Kyoto Protocol to the United Nations Framework Convention on Climate Change". Online at: http://unfccc.int/resource/docs/convkp/kpeng.pdf (accessed 10 January 2007).

Wackernagel, M. and W. Rees (1996). *Our Ecological Footprint: Reducing Human Impact on the Earth*. Gabriola Island, New Society Publishers.

Wambsganss, J. R. and B. Sanford (1996). "The problem with reporting pollution allowances". *Critical Perspectives on Accounting* **7**(6): 643–652.

Watson, C. (2009). "Forest carbon accounting: overview and principles". Addis Ababa, Ethiopia, UNDP-UNEP CDM Capacity Development Project for Eastern & Southern Africa. Online at: www.undp.org/climatechange/carbon-finance/Docs/Forest%20Carbon%20Accounting%20-%20Overview%20&%20Principles.pdf (accessed 8 January 2010).

World Business Council for Sustainable Development (WBCSD) and World Resources Institute (WRI) (2004). "The Greenhouse Gas Protocol: a corporate accounting and reporting standard (rev. edn)". Geneva and Washington, DC, World Business Council for Sustainable Development (WBCSD) and World Resources Institute (WRI).

Part III

COMMENSURATION

Part III of *The Making of Low Carbon Economies* concentrates on interdisciplinary scholarship about measurement, standardisation and calculation, using the overall theme of 'commensuration', defined as "... the transformation of different qualities into a common metric...." (Espeland and Stevens 1998: 314), in order to "... reduce and simplify disparate information into numbers that can be easily compared" (1998: 316). Commensuration is about creating equivalence and is seen as not just a technical project, but one that is deeply embedded within society. Scholars active in this field have researched topics as diverse as medicine and atmospheric science (Alonso and Starr 1984; Espeland and Stevens 1998; Bowker and Star 2000; Timmermans and Epstein 2010), and their research provides the following key insights:

- Commensuration is achieved through the creation and deployment of forms of measurement, calculation, classification and standards.
- Critical analysis of commensuration provides us with an understanding and appreciation of the day-to-day work involved in making diverse phenomena the same.
- Commensuration is largely taken for granted and is not usually 'seen'.
- A focus on things that do not fit into standard categories that have been constructed by society (termed 'incommensurables') yields the greatest conceptual insights.
- Acts of commensuration are sociotechnical and political.
- Measurement, classifications and standards are vital to markets and economies, because they provide a common calculative framework.
- Acts of commensuration – although they might appear stable – are always subject to change.

In this short introduction to Part III of this book, a number of key insights are explored that are revealed by scholarship on commensuration, which are especially pertinent for the study of low carbon economies. Commensuration is not an area of climate change research that has been given much attention (albeit with notable exceptions; see Levin and Espeland 2002; Lohmann 2009; MacKenzie 2009), and Part III of *The Making of Low Carbon Economies* aims to rectify this. Commensuration is important for climate change, particularly regarding carbon markets, because of how these markets have centred on the creation of carbon as a new and fungible commodity, in the process bringing together different greenhouse gas emission reductions across different locations, as part of an ambitious, highly technical and inherently political project to make them the same (MacKenzie 2009). An argument is thus made in Part III of this book for greater consideration of the role of standards and other acts of commensuration in the making of low carbon economies, building on the work of Higgins and Larner (2010), Barry (2006) and Dunn (2008), amongst others.

The first insight is that acts of commensuration (standards, forms of measurement and calculation) underlie and structure modern, everyday life. A central theme of commensuration scholarship is the invisibility of the multitude of standards that allow the modern world to function, with research demonstrating that once a standard or mode of calculation comes into being, it quickly becomes part of the routine backdrop of everyday life and is subsequently given little attention (Bowker and Star 2000; Busch 2000; Timmermans and Epstein 2010). As Timmermans and Epstein explain: "[s]tandards and standardization are such widespread and omnipresent features of modernity that, ironically, their precise sociological significance stands at risk of vanishing out of sight" (2010: 84). Bowker and Star (2000: 33), in their analysis of health services and race classifications and standards, similarly demonstrate how systems of measurement are typically paid little attention on a day-to-day basis: "Good, useable systems disappear almost by definition. The easier they are to use, the harder they are to see. As well, most of the time, the bigger they are, the harder they are to see".

From a Foucauldian governmentality perspective, one can conceive of acts of commensuration as a key element of the techniques and technologies that underpin government rationalities, by defining and prescribing "... visibilities, knowledge, techniques and practices, and identities ..." (Dean 1999: 23). Acts of commensuration frame issues and artefacts into technical programmes that can be managed (Higgins and Larner 2010), what Li refers to as "rendering technical" – a set of practices concerned with representing

> the domain to be governed as an intelligible field with specifiable limits and particular characteristics ... defining boundaries, rendering that within them visible, assembling information about that which is included and devising techniques to mobilize the forces and entities thus revealed.
>
> (Li 2007: 7, quoting Rose 1999: 52)

A key benefit of a governmentality lens for developing ideas about commensuration is its ability to bring together technologies and practices with discourse, viewing these elements of government as fundamentally connected in a two-way relationship. For instance, a governmentality approach shows us how carbon financial accounting governance and practice – the case study explored in Chapter 7 – can only be fully understood through examining the discursive rationale for the International Accounting Standards Board (IASB) in conjunction with its technical standards, as well as the framing and day-to-day operation of carbon markets. Applied to the case of carbon financial accounting, the practical, technical difficulties accountants are experiencing in classifying and managing emission allowances are thus linked to underlying ambiguities and tensions in the policy discourse and operation of carbon markets. Using a governmentality lens, financial accounting standards are transformed from something mundane and mostly hidden into a set of technologies with the potential ability to effect change in the discourse, framing and even the whole notion of carbon markets.

Carbon is a new and unusual commodity, meaning credible systems of measurement and calculation are especially important in developing low carbon economies. Scholarship on commensuration is highly relevant to the making of low carbon economies, because a large part of this enterprise – the work of bringing low carbon economies into being – has been to do with measuring and standardising the new commodity of carbon. Because carbon (in the sense discussed in this book – a commodity of greenhouse gas emission reductions) is an unusual and 'slippery' commodity (Bumpus 2011), it means that a stable calculative frame, defined as "the cognitive and material infrastructure that underlies economic calculations of value" (Callon *et al.* 2007: 33), is especially critical. It is through the "disciplinary technologies" of standards, rules, monitoring and auditing (Rabinow 1984; Lemke 2002; Foucault 2007; Mitchell 2008) that the commodity of carbon has been created and stabilised. We see this, for instance, in Chapter 6, where the creation of forest carbon markets has been, to date, almost entirely centred on debates about the measurement, reporting and verification of the carbon stored in forests.

Scholarship on commensuration provides a useful, detailed analytical frame for considering the processes involved in calculation, measurement and standard setting – in particular, the initial acts of classification required. Standardisation of new things (such as carbon) is usually done with reference to what has already been standardised – to the existing body of standards – but at the same time existing standards and classifications are always subject to revision: there is a constant tension between continuity and change. Low carbon economies cross numerous boundaries (political, financial, professional) and have manifested in different ways in different places, often challenging existing practices (Lovell and Navraj 2013). Standards and other acts of commensuration provide an empirical route into studying the effects of climate change in these diverse policy, technological and scientific arenas. Chapter 6, for example, demonstrates this tension between continuity and change in its discussion of the development of new standards of forest measurement and calculation, spurred on by the creation of low carbon markets for tropical forests (an initiative known as 'REDD+').

Climate change provides an interesting case through which to examine how pervasive modern systems of commensuration change, and how new ones develop in response to the emergence of this significant, new problem. For, as Chapters 6 and 7 demonstrate, there is nothing yet habitual about most low carbon practices, and they therefore stand in contrast to our established ways of doing. For issues such as climate change, relatively new on the scene, a focus on standards, measurement and modes of calculation is particularly pertinent, because new climate change policies, carbon commodities and ways of measuring greenhouse gas emissions must all somehow fit with acts of commensuration that already exist, be they in building design, medicine or accounting. Low carbon commensuration, therefore, is not just about constructing new carbon-based rules and procedures, but rather it has been a more complex process of fitting carbon into already existing modes of calculation and standards. The case of financial carbon accounting, explored in Chapter 7, demonstrates how this process of making climate change 'fit' is complex, uncertain and messy.

Acts of commensuration are sociotechnical (Tanaka and Busch 2003; Edwards 2004; Timmermans and Epstein 2010). They do not just involve people, but nonhuman things (objects, technologies, texts) too, as Timmermans and Epstein (2010: 84 [emphasis added]) explain with respect to standards: "Standardization consists of building a society around a standard with an implied script that *brings people and things together*". Acts of commensuration are the formal, tangible outcome of what is typically a fraught, politically charged and multi-dimensional process of detailed discussion and negotiation, as Edwards suggests in his insightful study of global meteorological standards: "Standards are socially constructed tools: they embody the outcomes of negotiations that are simultaneously technical, social and political in character" (Edwards 2004: 827). There are obvious overlaps here with the Part I theme of heterogeneous networks.

Social science scholarship about calculation (see, for example, Callon 2007; MacKenzie 2007) argues that far from being a rational, abstract process, calculation is in reality determined by a host of factors, including politics, technologies and types of expert knowledge. Standardisation is primarily viewed here as a way of "pacifying goods" (Çalışkan and Callon 2010: 5) so they are fit for markets – i.e. they remain stable (Callon 2007; Hardie and Mackenzie 2007). Standards, in this way, can be seen as a type of "market device" (see also the introduction to Part I) – a term coined to denote "the material and discursive assemblages that intervene in the construction of markets" (Muniesa *et al.* 2007: 2). It is evident that the work of standards and other acts of commensuration could be integral to market stabilisation, aiding stabilisation and thereby allowing markets to function. Indeed, Çalışkan and Callon (2010) in their enquiry into marketisation (the making of markets) specifically identify the process of standardisation of commodities as an important area of research: "Disentanglement is more stable ... when a commodity has undergone specific processes of standardization that transforms it into an entity described in both abstract and precise terms, certified and guaranteed by a series of textual and material devices" (2010: 8).

Standards and other acts of commensuration are devised by technical experts in the field – be it engineers, farmers, plumbers or doctors – and these experts have profound knowledge in the particular field that the standard relates to, but not typically beyond its boundaries (there are overlaps here with the concept of frames, see Part II). The narrow framing of standard setting expertise is the source of some of the problems encountered in the making of low carbon economies: a situation evident in the case studies of forest carbon (Chapter 6), where remote sensing expertise has been dominant, as well as in financial accounting, where accountants are shown to have limited knowledge of carbon markets (Chapter 7).

Another related insight (drawn from scholarship on sociotechnical transitions) is that change in sociotechnical systems (or economies) requires a vision for the whole system, including acts of commensuration, in order to flourish (see also the introduction to Part I). Mitchell (2008), for example, uses the case of the development of electricity networks in the United States to make the observation that what was required for these innovations to become widely used were new technical processes and forms of calculation that brought together and coordinated the many different aspects of electricity systems: "electrical, chemical, economic and social" (2008: 1117). Indeed, Mitchell (2008) goes further to argue that such processes and acts of calculation do not necessarily even have to be accurate; the important thing is that calculation works practically and is integrated within a system:

> ... successful calculative devices are not necessarily those that are most statistically complete or mathematically rigorous. They are those that make it possible to conceive of a network, or market, or national economy, or whatever is being designed, and assist in the practical work of bringing it into being.
>
> (2008: 1118)

It is demonstrated in several case studies on standards, on topics as diverse as the standardisation of rapeseed in China to medical classifications of types of death (Bowker and Star 2000; Busch 2000; Tanaka and Busch 2003), that, despite their objective, technical veneer, there is always ambiguity in how standards are made and operationalised, and hence there is ample room for politics, difficult choices and power struggles to surface (Barry 2005). This struggle in the process of making standards and new modes of calculation is evident, for example, in Chapter 6, which investigates the extension of carbon market rules, norms and practices of measurement to tropical forests. It is demonstrated that the policy ambition of accurately measuring forest carbon has become a dominant feature of the making of a low carbon economy for forests. The science of remote sensing is at the forefront of attempts to make forests manageable (i.e. measured) and therefore stable within international carbon markets. Chapter 6, in this way, explores the question of how carbon as a commodity is being integrated into existing long-standing rules, regulations and practices within different spheres of forest expertise – in particular, remote sensing. Such issues of commensuration are mostly invisible to the public, but are demonstrated to have a bearing on the overall effectiveness of forest carbon markets.

Critical analysis should focus on cases of incommensurables (things that do not fit), because they highlight unresolved tensions. A further insight garnered from scholarship on commensuration that is particularly relevant for analysis of the making of low carbon economies is the identification of so-called 'incommensurables'. Bowker and Star (2000) advocate concentrating critical analysis on incommensurable cases – defined as things that do not fit with existing acts of commensuration – because of their ability to highlight unresolved tensions. Once we begin to look in detail at climate mitigation activities, it is all too easy to spot instances of carbon as an incommensurable. In Chapter 7, for example, it is shown how and why emission allowances do not fit neatly under any existing financial accounting standard and are hence hard to classify, making them a type of incommensurable. The difficulties presented in accounting for emission allowances have their origins in the multiple potential uses of emission allowances: as a commodity, a currency, a financial instrument and so on (Bank of England 2009). As Espeland and Stevens (1998: 316) explain: "commensuration is noticed most when it creates relations among things that seem fundamentally different". What makes the financial accounting case study explored in Chapter 7 particularly noteworthy are the signs that the classification and measurement issues raised by emission allowances might have wider ramifications, prompting a rethink of the whole system of international financial accounting standards.

In summary, in this short introduction to Part III of the book, the concept of commensuration – creating equivalence – has been explained and its relevance to the making of low carbon economies outlined. The chapters within Part III apply these ideas to forest carbon markets (Chapter 6) and financial accounting (Chapter 7).

Bibliography

Alonso, W. and P. Starr (1984). *The Politics of Numbers: The Population of the United States in the 1980s*. New York, NY, Russell Sage Foundation.

Bank of England (2009). "Issue 116 – emission allowances: creating legal certainty". London, Financial Markets Law Committee.

Barry, A. (2005). "The anti-political economy". *The Technological Economy*. A. Barry and D. Slater (eds). London and New York, NY, Routledge: 84–100.

Barry, A. (2006). "Technological zones". *European Journal of Social Theory* **9**(2): 239–253.

Bowker, G. C. and S. L. Star (2000). *Sorting Things Out: Classification and its Consequences*. Cambridge, MA, The MIT Press.

Bumpus, A. G. (2011). "The matter of carbon: understanding the materiality of tCO_2e in carbon offsets". *Antipode* **43**(3): 612–638.

Busch, L. (2000). "The moral economy of grades and standards". *Journal of Rural Studies* **16**: 273–283.

Çalışkan, K. and M. Callon (2010). "Economization, part 2: a research programme for the study of markets". *Economy and Society* **39**(1): 1–32.

Callon, M. (2007). "What does it mean to say that economics is performative?" *Do Economists Make Markets? On the Performativity of Economics*. D. MacKenzie, F. Muniesa and L. Siu (eds). Princeton, NJ, Princeton University Press: 311–357.

Callon, M., Y. Millo and F. Muniesa (eds) (2007). *Market Devices*. Oxford, Blackwell Publishing.

Dean, M. (1999). *Governmentality: Power and Rule in Modern Society*. London, Sage Publications.

Dunn, E. C. (2008). "Standards and person-making in east central Europe". *Global Assemblages: Technology, Politics, and Ethics as Anthropological Problems*. A. Ong and S. J. Collier (eds). Oxford, Blackwell Publishing: 173–193.

Edwards, P. N. (2004). "'A vast machine': standards as social technology". *Science* **304**(5672): 827–828.

Espeland, W. N. and M. L. Stevens (1998). "Commensuration as a social process". *Annual Review of Sociology* **24**: 313–343.

Foucault, M. (2007). *Security, Territory, Population*. Basingstoke, Palgrave Macmillan.

Hardie, I. and D. Mackenzie (2007). "Assembling an economic actor: the *agencement* of a Hedge Fund". *The Sociological Review* **55**(1): 57–80.

Higgins, V. and W. Larner (2010). *Calculating the Social: Standards and the Reconfiguration of Governing*. Basingstoke, Palgrave Macmillan.

Lemke, T. (2002). "Foucault, governmentality, and critique". *Rethinking Marxism* **14**(3): 49–64.

Levin, P. and W. N. Espeland (2002). "Pollution futures: commensuration, commodification, and the market for air". *Organizations, Policy, and the Natural Environment: Institutional and Strategic Perspectives*. A. J. Hoffman and M. J. Ventresca (eds). Stanford, CA, Stanford University Press: 119–147.

Li, T. M. (2007). *The Will to Improve: Governmentality, Development, and the Practice of Politics*. Durham, NC, and London, Duke University Press.

Lohmann, L. (2009). "Toward a different debate in environmental accounting: the cases of carbon and cost-benefit". *Accounting, Organizations and Society* **34**(3): 499–534.

Lovell, H. and S. G. Navraj (2013). "Climate change and the professions: the unexpected places and spaces of carbon markets". *Transactions of the Institute of British Geographers* **38**(3): 512–516.

MacKenzie, D. (2007). "The material production of virtuality: innovation, cultural geography and facticity in derivatives markets". *Economy and Society* **36**(3): 355–376.

MacKenzie, D. (2009). "Making things the same: gases, emission rights and the politics of carbon markets". *Accounting, Organizations and Society* **34**(3–4): 440–455.

Mitchell, T. (2008). "Rethinking economy". *Geoforum* **39**: 1116–1121.

Muniesa, F., Y. Millo and M. Callon (2007). "An introduction to market devices". *Market Devices*. M. Callon, Y. Millo and F. Muniesa (eds). Oxford, Blackwell Publishing/The Sociological Review: 1–12.

Rabinow, P. (ed.) (1984). *The Foucault Reader: An Introduction to Foucault's Thought*. St Ives, Clays Ltd.

Tanaka, K. and L. Busch (2003). "Standardization as a means for globalizing a commodity: the case of rapeseed in China". *Rural Sociology* **68**(1): 25–45.

Timmermans, S. and S. Epstein (2010). "A world of standards but not a standard world: toward a sociology of standards and standardization". *Annual Review of Sociology* **36**: 69–89.

6

MEASURING AND STANDARDISING FOREST CARBON FOR REDD+

Key findings

- This chapter explores the discourse and technologies of forest carbon 'measurement, reporting and verification' (MRV), using Foucault's ideas about governmentality.
- The notion of governmentality complements scholarship on commensuration, particularly through its close attention to discourse and politics.
- Methods for measurement, reporting and verifying the carbon stored in forests, and standards that determine these methods, are crucial governmentality techniques (see Dean 1999) that have allowed forest carbon markets to come into being.
- Forest carbon MRV is difficult to do and has taken up significant resources (expertise, time, money) in international climate change governance.
- An excessive focus on forest carbon MRV has, however, necessarily limited the scope of forest carbon markets – things and institutions are excluded, because they cannot be "rendered technical" (see Li 2007: 7) through MRV.

Introduction

My first introduction to the measurement of carbon stored in forests was a presentation by a colleague of mine at Edinburgh University – Professor Mat Williams – at an International Conference on the African Miombo woodlands, held in Edinburgh in 2008 (Williams 2008). In this presentation he showed a picture of a sizeable backhoe vehicle digging up tree roots (see Plate 6.1). It looked like the forest was being cleared for a new development or building, but what the picture shows in actual fact is the considerable amount of destruction required in order to fully measure the carbon stored in a tree. This is a lot of work, particularly if one takes into account, as Mat explained, the difficulty of sourcing a backhoe in this remote part of Africa and then transporting it to the field site. It

Plate 6.1 A backhoe digger excavating tree roots in order to measure the total amount of carbon stored in a sample of trees, Miombo woodlands, Mozambique (source: Dr Casey Ryan, University of Edinburgh).

piqued my interest in how carbon is measured in forests and whether there are easier ways of doing it. This led me to the science and practice of remote sensing: certainly an easier method, but one that also has limitations, as this chapter explores.

Accurate measurement of forest carbon has become a key element of attempts to integrate forests more fully into international climate policy through markets. There is widespread agreement about the value of including forests in a post-Kyoto United Nations (UN) treaty, as tropical deforestation and other land-use change constitutes an estimated 8 per cent of global greenhouse gas emissions (Global Carbon Project 2013). The basic idea of the current UN Framework Convention on Climate Change (UNFCCC) policy initiative on forests, termed 'Reducing Emissions from Deforestation and Forest Degradation' (and widely referred to as 'REDD+')[1] is a simple one: to financially reward developing countries who protect their forests. However, the translation of this overarching policy objective into a workable set of policy guidelines, standards and practices has proved much more difficult, and work is still ongoing to decide upon key issues. In this chapter, ideas about commensuration are assessed using the case of international efforts to develop markets for the carbon stored in tropical forests.

This chapter draws on the conceptual lens of Foucault's notion of governmentality in combination with theories of commensuration (see the introduction to Part III).

The UNFCCC has agreed that the 2003 Intergovernmental Panel on Climate Change (IPCC)[2] Good Practice Guidance (GPG) should form the basis for measuring forest carbon under REDD+ (see UNFCCC 2007: Decision 2/CP.13(6)). However, the IPCC GPG alone does not provide sufficient detail to operationalise REDD+ and make it work. A huge amount of activity has therefore emerged, concentrating on defining the precise practices and technologies of forest carbon 'Measurement, Reporting and Verification' (known within REDD+ and wider carbon market networks as 'MRV'), resting on IPCC guidelines, but fleshing them out through the provision of best practice examples, guidelines, standards, protocols and manuals. It is this 'IPCC plus' area of activity that forms the core empirical focus of this chapter: who is doing this work, and why? How have scientists been involved? How are new MRV practices and technologies shaping forest carbon markets? The focus of this chapter is on the activity of a range of organisations (mostly either NGOs or science institutions) in defining forest carbon measurement standards aimed at operationalising the IPCC methodological guidance (Penman *et al.* 2003).

Foucault's concept of governmentality, which brings together analysis of discourse ('rationalities') and instruments, practices and technologies ('techniques') (Foucault 1991; Dean 1999) provides a means to usefully explore such issues of measurement, calculation and standardisation in forest carbon markets. The implication of the conceptual linking of 'rationalities' with 'techniques' in governmentality for the study of low carbon markets and economies is that the technical capability to provide market relevant information on climate mitigation (carbon statistics and data) is integral to the successful framing and operation of low carbon economies. But also, as leading scholars have observed (Luke 1999; Lovbrand and Stripple 2010), the 'measure to manage' rationality that pervades carbon markets and low carbon economies in itself limits climate mitigation action to a relatively narrow band of things and activities that are deemed to be cost-effective and can be measured and monitored. As the case of forest carbon market MRV demonstrates, an economic framing of climate change mitigation places stringent boundaries on what is viewed as credible, acceptable and workable.

This chapter is based on a number of different strands of empirical research undertaken as part of a three-year Nuffield Foundation Fellowship (2008–2011; held jointly with Professor Donald MacKenzie, Edinburgh University), including: an online survey of authors (44) of a lead handbook for forest carbon measurement (the "Sourcebook" published by the international organisation Global Observation of Forest and Land Cover Dynamics (GOFC-GOLD)) in late 2011; a follow-up online survey of users of the GOFC-GOLD "Sourcebook" (116) in August 2012; participant observation and work shadowing of a remote sensing scientist at the University of Edinburgh for two months (Summer 2010); semi-structured interviews (21) (2010–2012) with leading international forest scientists in academia, NGOs, the UN and the private sector; and attendance at a number of REDD+ and forestry seminars and conferences.[3] This primary empirical research has been complemented by a policy 'grey' literature review of REDD+ documents related to MRV and an academic science literature review of the use and application of allometric equations (with respect to both timber and carbon markets; see Lovell and MacKenzie 2014).

The structure of the chapter is as follows. First, the international forest carbon policy context is briefly explained, highlighting recent decisions, as well as providing some historical perspective to contemporary debates. Second, in the main empirical section of the chapter, the discourse and techniques of forest carbon MRV are described and analysed using the concept of governmentality, combined with STS ideas about commensuration. In conclusion, the role of scientists and other experts in acts of commensuration is commented upon and the implications of forests being living, rather unstable, policy and economic actors is discussed.

A brief introduction to forest carbon markets

The core objective of REDD+ is to financially compensate developing countries (termed 'non-Annex One' countries, under the UNFCCC) that prevent deforestation and degradation of their carbon-rich tropical forests. It was the 2007 Bali Action Plan at the UNFCCC Conference of the Parties (COP) 13 that marked the formal agreement to reinvigorate the role of forests within the UNFCCC, stating that a comprehensive approach to mitigating climate change should include: "Policy approaches and positive incentives on issues relating to Reducing Emissions from Deforestation and Forest Degradation in developing countries; and the role of conservation, sustainable management of forests and enhancement of forest carbon stocks in developing countries". (UNFCCC 2007: Decision 1/CP.13, para 1b iii). Whilst much has been researched about REDD+, including in-depth analysis of its politics (see, for example, Boyd *et al.* 2007), the focus of Chapter 6 is limited to the discourse and techniques of MRV. MRV is an interesting focus of social science enquiry, because it is viewed as the cornerstone of carbon markets: accurate measurement and reporting of carbon is seen as central to making international carbon markets work (see Chapter 5; also Lovbrand and Stripple 2010), and this is equally true for the newer subset of forest carbon markets (Grainger 2009). For example, one recent UN REDD Programme publication notes that:

> . . . the potential benefits [of REDD+] for non-Annex One parties *will be based on results that must be measured, reported and verified. The precision of these results therefore has a major impact on potential financial compensation*, and the capacity to measure forest carbon stocks is thus of increasing importance for countries who plan to contribute to mitigating climate change through their forest activities.
>
> (Picard *et al.* 2012: 17 [emphasis added])

One issue it is important to appreciate with regard to REDD+ forest carbon MRV is that Annex One countries (developed countries of the Global North) under the UNFCCC have been measuring and reporting on carbon in their forests (and other carbon 'pools') for some years, as signatories to both the Kyoto Protocol and UNFCCC. Annual reports on greenhouse gas sources and sinks are required by the UNFCCC (under Article 4), with further, more detailed carbon pool measurements required for countries that are signatories to the Kyoto Protocol (see Watson 2009: 8, for a helpful summary of these complex rules). This means that developed countries have built up considerable experience and expertise on forest carbon measurement. In contrast, non-Annex One developing countries – the focus of REDD+ – have not, in the main, been reporting to the UNFCCC on their national inventories of emissions and sinks (including forests), as they have not been required by the UNFCCC to do so on an annual basis. A study of ninety-nine tropical forest countries found that less than 20 per cent of them have submitted a complete greenhouse gas inventory to the UNFCCC, and only three out of the ninety-nine non-Annex One countries have 'very good' capacity for monitoring forest carbon (Herold 2009).

Much could be said about the history of the role of forests in climate mitigation – a process that has largely taken place at an international level, through the UNFCCC (for excellent detailed discussions, see Fogel 2005; Backstrand and Lovbrand 2006; Stephan 2012). In short, forests have had a rather controversial role, largely stemming from unease on the part of developed countries about the implications of including tropical forests in an international agreement. Under the CDM (see Chapter 2), some forestry projects were allowed, but, contrary to expectations, in practice only a handful of projects came to fruition (because these projects were quickly discovered to be less profitable, more complicated and risky than other types of CDM project; see Fogel 2005). It was hoped, therefore, with the initial REDD agreement at the UNFCCC COP-13 meeting in Bali that a new era of forest protection under the UNFCCC had been born. Since the 2007 Bali COP, more detail has followed at each COP and interim UNFCCC meetings (notably, meetings of the Subsidiary Body on Scientific and Technical Advice (SBSTA)), including how to measure carbon in forests. But progress has been slow – held up in part by problems with the overall climate negotiations –

leading to the rather curious current situation characterised by extensive, detailed debate and activity concerning how to do REDD+ (i.e. how to operationalise it), yet without a final REDD+ policy agreement being in place.

The central role of MRV within forest carbon markets

From this brief introduction to REDD+, it is evident that knowledge and experience of forest carbon measurement in non-Annex One countries is patchy at best. This 'capacity gap' is further compounded by anticipated differences in the degree of accuracy required to measure the carbon stock of forests at a national scale (as for Annex One developed countries), compared with measuring carbon in specific forest projects (as is under discussion as part of REDD+, for non-Annex One developing countries) and, crucially, in a situation where financial compensation is to be involved. Much policy discussion about REDD+ has therefore centred on the issue of achieving so-called 'compliance-grade' MRV – namely, MRV that is sufficiently accurate to reassure investors. This is a policy discussion in which scientists have played an important role, including convincing and reassuring policymakers that implementing REDD+ policy (however it may be finalised) is technically possible, as the introduction to the popular "Little REDD+ book", produced by the Global Canopy Programme (a forest research organisation), summarises:

> Agreement on REDD is within reach. The spread of new technologies such as satellite monitoring is overcoming some long-standing technical barriers. Collaboration by scientists, economists and policy makers at the UNFCCC, IPCC and other forums, is helping to clarify outstanding methodological issues.
>
> (Global Canopy Programme 2009: 8)

As the Global Canopy Programme suggests, various organisations – intergovernmental and non-governmental organisations, private sector, science/research centres – have been busy developing methodologies and guidance for countries in preparation for REDD+.

But even before the fresh policy impetus of REDD+, forest carbon measurement guidance was issued by the Intergovernmental Panel on Climate Change (IPCC) in response to requests from Annex One countries (who were measuring forest carbon for their annual reports, as previously outlined). In 2003, the "IPCC Good Practice Guidance (GPG) for land use, land-use change and forestry" was published (Penman *et al.* 2003), which then subsequently merged with agricultural guidance to become "Volume 4 Agriculture, Forestry and Other Land Use (Guidelines)" (AFOLU) – of the IPCC for "Guidelines for national greenhouse gas inventories", published in 2006 (Eggleston *et al.* 2006). As mentioned, it is currently agreed that the 2003 IPCC GPG should form the basis for measuring forest carbon under REDD+ (see UNFCCC 2007: Decision 2/CP.13(6)). However, the IPCC GPG alone does not provide sufficient detail to operationalise REDD+ and make it work, as evidence from the research of forest and remote sensing scientists discussed later reveals. A report for UNEP summarises the situation as follows:

> As the leading intergovernmental scientific body for the assessment of climate change, the IPCC is in the foremost position to form the basis of any future accounting guidance under an international climate change convention. If a mechanism for reduced emissions from degradation and deforestation (REDD) emerges . . . *it is likely to build on the principles and good practice guidance already established in IPCC guides.*
>
> (Watson 2009: 26 [emphasis added])

The result of this absence of sufficient, comprehensive UNFCCC standards for REDD+ MRV is that several organisations have been actively developing new standards, united by a shared view of the central role of MRV in operationalising forest carbon policy. For the remainder of this chapter, the discussion concentrates on how we might best understand the development of these new MRV standards, drawing on Foucault's concept of governmentality, as well as theories of commensuration (see the introduction to Part III; also Barry 2006; Timmermans and Epstein 2010).

Governmentality – understanding the consensus around forest carbon MRV

In common with other multi-level environmental governance debates (Bulkeley and Betsill 2003), the discussions around REDD+ are complex – policymakers and others shaping the policy process are dealing with myriad issues of forest science and policy operating at a variety of spatial scales. As is typical within new arenas of policymaking, language and practices are particularly important, because in numerous subtle ways they shape choices about rules and norms, thereby framing the policy debate and propelling it in a particular direction (Dryzek 1997; Owens and Cowell 2002; see also Part II of this volume). In other words, how a policy problem like REDD+ is first discussed and operationalised affects how it is conceived and thus the choice and acceptability of subsequent solutions (Rein and Schon 1993; Laws and Rein 2003). The distinctiveness and applicability of governmentality as a theoretical approach here is its bringing together of discourse with 'techniques' (practices and technologies; see Dean

1999). Further, it is a theory that explicitly embraces a range of different types of policy actor: both state and non-state (see, for example, Bryant 2002), and thus it is highly applicable to the hybrid mix of organisations involved in forest carbon MRV (see Table 6.1). In these two ways, governmentality complements theories of commensuration, which have overall been less attentive to issues of discourse, as well as the diverse types of organisation involved in commensuration.

The discourse of forest carbon MRV

The "storyline" (see Hajer 1995; also Part II) of the dominant discourse of forest carbon MRV is that in order for developing countries' forests to be successfully incorporated into the UNFCCC under REDD+, rigorous 'compliance-grade' systems of MRV must be developed and systematically implemented at a national level. These national systems need to measure both changes in forest cover (i.e. deforestation, reforestation – the quantity of forest), as well as the quantity of carbon per hectare – the carbon stock or concentration (see UNFCCC 2009: Copenhagen Accord: Decision 4/CP.15). The MRV storyline has constructed global tropical forests as wild, unmanageable and unquantified: untamed forests that need to be brought to order and under control through systematic measurement by nation states (see Lovbrand and Stripple 2010 for related discussions about climate MRV). Forest carbon MRV is about monitoring, data gathering and close surveillance of forests.

The dominance of the framing of forest carbon policy around MRV is key to understanding the vast amount of work devoted to devising a workable, robust, global system of measurement for carbon in forests, which has, in turn, shaped international climate negotiations. Taken together, this discursive framing and its associated set of practices (about which, more later) constitute a governmentality of MRV. Acknowledging this allows much of what is taking place in forest and climate governance to be better understood, with the MRV governmentality acting to "shape the realm of the possible", as Lovbrand and Stripple (2010: 21) convincingly argue in their analysis of space-based monitoring of climate change:

> To understand the current shape of climate politics and possible climate policy futures, we must pay more attention to how practices like MRV have enabled the climate as an administrative domain and how they shape the realm of the possible.

But despite the daunting size of the task – the unwieldy, huge "administrative domain" of tropical forests – there is an optimism to the MRV storyline: a belief in the UNFCCC position that one can 'know' the forest and translate it into policy through a systematic process of measuring, quantifying, creating data infrastructure, verifying and so on. It is an optimistic 'measure and manage' discourse, as an extract from an interview transcript with Dr James Baker, Director of Clinton Climate Initiative's Global Carbon Measurement Program, illustrates:

> Why is measurement important? The short answer is: You can't value what you can't measure. . . . Only by measuring forest carbon reliably can countries access the carbon markets and other sources of investment capital that would make forest preservation economically viable. *Accurate information truly underpins everything communities, governments, and charities are doing to halt tropical deforestation.*
>
> (Clinton Climate Initiative 2012 [emphasis added])

A surprisingly diverse range of organisations participate in the MRV storyline (see Table 6.1), ranging from community-based forest groups to national foresters, scientists and financial organisations.

With the high degree of consensus around the MRV storyline (i.e. with virtually no-one saying 'don't bother to accurately measure carbon in forests'), attention has been heavily concentrated on how to measure forest carbon, not why. How has this situation emerged? Why has a MRV governmentality come to dominate? Part of the answer to these questions relates to existing forestry and climate expert knowledges and practices that have shaped the direction of forest carbon policy. Several spheres of expertise, which pre-date the emergence of REDD+ – timber management, forest ecology, global land-use change studies, global carbon budget science, remote sensing and community forest management – have all been an important influence on the direction of REDD+ policy and practice in different ways. Whilst there is not scope within this chapter to examine all of these bodies of knowledge and expertise, I focus on arguably the most influential one – that of remote sensing – which has come to be particularly important within the MRV governmentality. Remote sensing has positioned itself as providing the technological capability, via a network of sophisticated satellite sensors, to measure forest carbon with high frequency at a global scale – a large-scale high-tech surveillance system that promises high quality "eco-knowledge" (Luke 1995). However, remote sensing is less convincing at providing tree-level data and information on changes below the forest canopy (forest degradation and conservation), meaning that other, more local-scale MRV practices (forest ecology, forest community monitoring) are also jostling for attention within the MRV governmentality. For the remainder of this chapter, I, first, explore in more detail the relationship of remote sensing to the MRV governmentality and, second, discuss the role of forest carbon standards in attempting to bring together different types of MRV expertise and practices, or 'techniques', including ground-based and space-based data; expert and lay knowledges.

Table 6.1 Examples of use of the Measurement, Reporting and Verification (MRV) storyline by diverse organisations

Organisation	*Position on forest carbon MRV*	*Reference*
Centre for International Forestry Research (CIFOR)	"Full national participation in a global REDD+ system requires a far better MRV system than currently exists, and there is a huge capacity gap." (p. xiv) "Better MRV systems are needed to develop performance-based payment systems. One concern is that high transaction costs (e.g. for forest carbon monitoring) prohibit the inclusion of local communities . . . Recent work on community monitoring demonstrates that their costs can be substantially lower compared to professional surveys, and the accuracy is relatively good." (p. xiv)	Angelson, A. (ed.) (2009). "Realising REDD+: national strategy and policy options." www.cifor.org/publications/pdf_files/Books/BAngelsen0902.pdf
The Meridian Institute	"A GHG-based instrument that rewards REDD on the basis of quantified emission reductions and/or removal enhancements requires agreement on standards for MRV." (p. ix)	REDD Options Assessment Report (OAR) (2009). Online at: www.redd-oar.org/links/REDD-OAR_en.pdf
The World Bank (Forest Carbon Partnership Facility)	"The UNFCCC has made it clear that the eligibility of countries for REDD-plus financing within global compliance markets or non-market compliance mechanisms will be contingent on their having compliance-grade MRV and accounting of emissions and removals." (Baker *et al.* 2010: 35) "Addressing methodological issues such as Reference level and measurement, reporting and verification (MRV) is a key entry requirement for REDD+ programs." (FCPF 2011: 26)	www.forestcarbonpartnership.org/fcp/sites/forestcarbonpartnership.org/files/Documents/PDF/Oct2011/FCPF_Carbon_AR_FINAL_10_3.pdf

UN REDD Programme	"Monitoring systems that allow for credible measurement, reporting and verification of REDD+ activities are among the most critical elements for the successful implementation of any REDD+ mechanism."	www.un-redd.org/UNREDDProgramme/InternationalSupport/MeasurementReportingandVerification/tabid/1050/language/en-US/Default.aspx
Global Canopy Programme	"As part of a future UN climate treaty, REDD+ could pay forest-owning nations who protect their forests. In return, these nations will have to measure, report and verify (MRV) the effectiveness of their actions, both in terms of the carbon retained in their forests and the involvement of indigenous peoples and local communities in the process. In addition, on-the-ground implementation of REDD+ will require robust but cost-effective monitoring to ensure that funds are being well spent. Communities should have a central role in developing and implementing these MRV and Monitoring systems."	www.globalcanopy.org/projects/community-mrv
Group on Earth Observations (GEO) Global Forest Observations Initiative (GFOI)	"Forest monitoring is receiving particular attention because of the potential inclusion of forests in future international climate change agreements. Robust measurement, reporting and verification (MRV) systems are a fundamental forerunner to any such inclusion." (p. 2) "To operate efficiently and sustainably, national forest monitoring systems require a continuous, timely and affordable supply of observations." (p. 1)	Geo Forest Carbon Tracking Report on the Concept Phase for developing a Global Forest Monitoring Network (2010) & GEO Global Forest Observations Initiative (GFOI) Implementation Plan (2011)

The techniques of forest carbon MRV: Part I – remote sensing science

For the remote sensing science community, MRV is a set of practices carried out from space using a range of satellite sensors in order to capture relevant forest data at a global scale. Remote sensing is presented as the only set of technologies that matches the scale of the task – i.e. global, accurate, wall to wall coverage of tropical forests, for example:

> . . . satellite data . . . play an essential role in delivering the necessary time-series global, annual, high resolution wall to wall forest information products which developed and developing countries will need as part of their MRV systems.
>
> (Baker *et al.* 2010: 42)

Similarly, "the scale of sampling must match the scale of the subject to be measured, in this case the carbon stocks of tropical forests" (UNFCCC 2009: 28, para 93). Remote sensing science invokes forest carbon MRV as necessitating an ambitious, high cost and high-tech approach involving the development of a whole new global data infrastructure and monitoring system. Parallels are frequently drawn to the establishment of the World Meteorological Organisation (WMO), in terms of the amount of data required and the size and credibility of the international organisation needed to manage it:

> Inter-governmental organizations, in particular the UN system, have a key role in operationalising a [forest carbon] data supply infrastructure. The experience of weather-related satellite data and the role of the WMO in providing the infrastructure is an important and successful example.
>
> (UN REDD Programme 2008: 5, para 7)

Also, "compared to the global networks which are the basis for meteorological and oceanographic monitoring, the international co-ordination and coherence of terrestrial observation and reporting for forests is much less advanced" (GEO Forest Carbon Tracking 2010: 2). The obvious appeal of remote sensing to policymakers is the potential for a technically reliable global system of MRV, which, although costly to implement, would then in theory deliver a long-term robust and tangible solution to the problem of gaining accurate data on forest carbon. As noted, the science of remote sensing provides an appealing body of knowledge and expertise to REDD+ policymakers wishing to make commensurate the carbon stored in tropical forests with other types of carbon within international carbon markets (see Chapters 2 and 5).

One international organisation – the Global Observation of Forest and Land Cover Dynamics (GOFC-GOLD) – provides an illustrative example of how and why remote sensing science has become key within the MRV storyline. GOFC-GOLD is a UN intergovernmental organisation (a Panel of the larger Global Terrestrial Observing System (GTOS)) with a strong earth observation and remote sensing remit. Its overall objective is to "... improve the quality and availability of observations of forests and land cover at regional and global scales and to produce useful, timely and validated information products from these data for a wide variety of users" (GOFC-GOLD 2012). The GOFC-GOLD good practice guide or "Sourcebook" (hereafter referred to as the GOFC-GOLD Sourcebook) was developed by its REDD Working Group, and has been driven by UNFCCC REDD+ policy, provides a detailed field guide or 'how to' book – a set of "agreed upon rules" (Timmermans and Epstein 2010: 71); with its stated objective of complementing IPCC guidelines, "... the Sourcebook aims to provide additional explanation, clarification, and methodologies to support REDD early actions and readiness mechanisms" (GOFC-GOLD 2009: 2).

The discussion here is based on empirical findings of a survey conducted of GOFC-GOLD authors (44; response rate 65 per cent) in late 2011, as well as a related set of interviews (14). Survey participants were asked several questions, including how they came to be involved in the GOFC-GOLD Sourcebook, how they account for its influence and their views on the possibility of scientists developing forest carbon measurement standards that meet policymakers' expectations. The organisation GOFC-GOLD and its Sourcebook were selected for investigation after an initial two-month period of enquiry in mid-2010 on forest carbon MRV, because the GOFC-GOLD Sourcebook has become one of the most well-referenced, science-based guidelines for REDD+ (UNFCCC 2009; Baker *et al.* 2010; GEO Forest Carbon Tracking 2010), as well as being widely cited in preliminary interviews and at conferences as a 'state of the art' science text. Further, the Sourcebook has a strong focus on new, emerging technology (it is updated each year in time for the annual COP intergovernmental negotiations under the UNFCCC), with a particular emphasis on remote sensing, which enabled further investigation into how this branch of science knowledge and expertise has been applied and translated into policy. Table 6.2 provides a summary of the main findings of the survey of GOFC-GOLD Sourcebook authors about the role of knowledge and their views on how the GOFC-GOLD Sourcebook has influenced the policy process through the provision of expert knowledge.

For scientists who study forests, one might expect them to be mostly quite cautious about the ease and potential of getting to 'compliance-grade' MRV of forests, and results from the GOFC-GOLD author survey to some extent back this up. As the selection of survey findings presented in Table 6.2 illustrate, there

Table 6.2 GOFC-GOLD author survey findings

Finding	*Illustrative GOFC-GOLD author survey responses*
Methodological feasibility of REDD+: The Sourcebook has been important in giving policymakers confidence that REDD+ is technically possible.	"The Sourcebook passed the following key message to policy makers: monitoring and reporting emissions from deforestation and forest degradation in tropical countries is technically feasible with acceptable accuracy for a REDD+ mechanism." (Respondent 4) "When policy makers were concerned on the feasibility of REDD, the GOFC-GOLD Sourcebook sent a positive message, which I think was influential on REDD+ negotiations." (Respondent 13) "The Sourcebook helped in solving the questions over the current technical capacity to measure forest carbon stock changes in developing counties which were some of the main concerns at the beginning of the REDD+ negotiations." (Respondent 26)
Relationship with IPCC: The focus of the Sourcebook has been on translating IPCC Guidelines, providing detailed country case studies and examples to show what achieving REDD+ might mean in practice.	"The Sourcebook makes more user-friendly methods described in the IPCC guidelines providing additional information to guide stakeholders in implementing those methods ensuring that the principles on which IPCC methods have been built (i.e. transparency, completeness, consistency, comparability and accuracy) are respected in monitoring forest carbon stocks and carbon stocks changes." (Respondent 14) "[The] Sourcebook is like a guidebook developed on the basis of IPCC Guidelines with case studies and examples for those who would like to learn about and implement MRV and REDD+ issues [to do with the] carbon stock of forest ecosystems." (Respondent 21)
Scepticism about policymakers' knowledge and comprehension of forest carbon measurement – for example, regarding specific areas of knowledge to do with classifying different types of forest carbon MRV, but also more broadly in terms of a lack of understanding of the scientific process.	"One point which is essential is to separate out is the issue of assessing reduced deforestation (including conservation) and issues of assessing [forest] degradation (including forest enhancement and sustainable management of forest), since the first 2 are primarily based on area change and the last 3 primarily on stock change. . . . *Most policy makers hardly understand the difference.*" (Respondent 1) "The way forest carbon is measured will influence the results – no two methods will yield exactly identical results and so if international policy makers are waiting for the scientific community to settle on one agreed upon approach for measurement then I am not optimistic." (Respondent 11)

is doubt amongst some in the science community that forests can ever be subject to a workable system of MRV and be made governable within international climate policy, as one survey respondent comments:

> The GOFC-GOLD Sourcebook has definitely influenced the negotiators and bureaucrats in REDD+ – it has told them that "yes, we can" measure forest change globally, but I'm not sure I agree on that. In that sense the Sourcebook has been a bit misleading.
>
> (GOFC-GOLD survey, Respondent 6, 2011)

But it is a mixed picture, with other remote sensing scientists being much more certain about the potential of remote sensing to provide an accurate, global measure of forest carbon. It is clear that the techniques that populate the forest carbon governmentality of MRV are not yet fixed; indeed, if remote sensing is the forerunner of these techniques, then it signals that the ambitious discourse of MRV might be somewhat ahead of its technological capabilities.

The techniques of forest carbon MRV: Part II – standards

One important way to discern the key governance techniques and practices in any particular policy sector is to assess the standards (the rules, principles and 'how to' guides) that are used to operationalise government. It is through standards and guidelines that existing forest carbon expertise and knowledge is being translated into REDD+ MRV policy and practice. It is also through standards that we observe attempts to make forest carbon markets commensurate with other carbon markets. In the case of forest carbon MRV, standards have become a key avenue for translating knowledge and expertise into the policy realm, both as a route to enacting that expertise and also in gaining legitimacy and credibility for the organisations involved. Table 6.3 highlights a small selection of the many standards, protocols and 'how to' guides that populate the field of forest carbon MRV.

As the data in Table 6.3 demonstrates, in this policy area a diverse range of organisations are active: state and non-state; expert science and community-based. As noted, for forest carbon MRV the 2003 "IPCC GPG for land use, land-use change and forests" is, in effect, the baseline standard. The IPCC GPG provides international principles and guidance for measuring and reporting on carbon stock changes and emissions for different land-use categories and carbon pools, including forests. IPCC Monitoring Standards (Tier 1, 2, 3) set out within the IPCC GPG stipulate different levels of accuracy for forest carbon measurement: Tier 1 is the most basic, using standardised continent-scale data (averages for six vegetation types); Tier 2 has some country-level data, based on forest

Table 6.3 Examples of forest carbon MRV standards

Name of organisation	*Type of organisation*	*Forest carbon MRV standard*	*Comment*	*Key references*
Centre for International Forestry Research (CIFOR)	Research-orientated NGO	Forest Carbon Database (2010)	The aim of the CIFOR Forest Carbon Database is to provide open access information for scientists and other practitioners, to avoid duplication and allow comparison of forest carbon data, based on IPCC guidelines (see p. 11 of the "User Guide" (online at: www.cifor.org/publications/pdf_files/ForestCarbonDatabase_web.pdf?_ga=1.259599370.1025258491.140964867).	www.cifor.org/publications/pdf_files/Books/BKurnianto1001.pdf; http://carbonstock.cifor.cgiar.org/
Intergovernmental Panel on Climate Change (IPCC)	Intergovernmental UN science agency; under auspices of UNFCCC	2003 IPCC Good Practice Guidance (GPG) for Land Use, Land-Use Change and Forestry (LULUCF)	Providing international principles and guidance for measuring and reporting on carbon stock changes and emissions. IPCC was invited to do this GPG under the Marrakesh Accords.	IPCC GPG for LULUCF (2003). Online at: www.ipcc-nggip.iges.or.jp/public/gpglulucf/gpglulucf_contents.html IPCC (2006) for ALOFU
Clinton Climate Initiative (CCI)	NGO	Carbon Measurement Collaborative	"CCI helps countries comply with international MRV standards, building a credible database on which to advance international agreements on deforestation. CCI also helps to ensure that critical, national-scale satellite information on forests is delivered regularly and routinely to countries that depend on it." (CCI website (accessed 14 January 2011))	www.clintonfoundation.org/what-we-do/clinton-climate-initiative/forestry/measuring-carbon

World Bank Forest Carbon Partnership Facility (FCPF)	Intergovernmental, UN	Country Readiness Preparation Proposals		https://www.forestcarbonpartnership.org/
Group on Earth Observations (GEO) Global Forest Observation Initiative	Intergovernmental, voluntary partnership of governments and international organisations	In preparation (see *Implementation Plan*), 2014+ will be developing methods and protocols		Group on Earth Observations (GEO) *Global Forest Observation Initiative Implementation Plan* (2011)
GOFC-GOLD	International state-based organisation	GOFC-GOLD Sourcebook – updated at each COP, provides information on methods and procedures for MRV	(See main empirical discussion).	
Climate Community and Biodiversity Alliance (CCBA) standards	Partnership of international NGOs, inc. CARE, Rainforest Alliance, Conservation International	Carbon standard for voluntary market land-based projects that encompasses not just carbon reductions, but other positive project benefits (poverty alleviation, etc.)	The goal of CCB standards is to "catalyze a robust carbon market for multiple-benefit forest carbon projects" (CCB standards, 2nd edn, p. 4). It rests on IPCC guidelines for AFOLU (2006).	www.climate-standards.org/
Verified Carbon Standard (VCS)	Founded by carbon market organisations and NGOs	Leading voluntary carbon market standard	Often used in conjunction with CCB standards (as CCBA does not actually issue credits). Has a separate AFOLU Expert Committee.	www.v-c-s.org/develop-project/agriculture-forestry-projects

inventories; and Tier 3 is the most robust standard, based on individual sample plots. So whilst the standards in Table 6.3 rest on the IPCC GPG, there remain, as noted, a host of other secondary activities not defined by the IPCC that have not yet been standardised and that must be agreed upon in order to develop a workable system of MRV for REDD+. It is in this fast-moving policy space – stemming from IPCC GPG and UNFCCC policy decisions and beyond – that competing techniques of the MRV governmentality are jostling.

Remote sensing science, as discussed, is the knowledge domain and set of techniques most closely associated with forest carbon MRV, but there are other competing techniques – in particular, ground-based local community forest monitoring (i.e. MRV techniques carried out by forest communities themselves, based on their local knowledge and capabilities), with its approach and methodologies derived from the discipline of forest ecology (see Lovell 2014). According to the UNFCCC, REDD+ MRV must include both types of MRV techniques: ground-based and remote sensing. Indeed, the UNFCCC has been quite explicit in this regard (see UNFCCC 2009: Decision 4/CP.15, 1d(i)). The challenge, and to date largely unresolved difficulty, where tensions have emerged, is how to *integrate* these different types of MRV techniques, each with their own methods and ideas, but retain a shared commitment to the overall objective of forest carbon MRV. The proliferation of standards on forest carbon MRV is, in large part, in response to the continuing uncertainty: the acts of commensuration required to integrate and stabilise forests within international carbon markets are not yet agreed upon.

One useful way to understand how these different types of forest carbon MRV expertise are jostling together is to examine the properties of the forests and trees themselves and their capacity to be measured. As outlined in the introduction to Part III, acts of commensuration are always necessarily sociotechnical, involving people as well as non-human entities (objects, technologies, texts) (Tanaka and Busch 2003; Edwards 2004; Timmermans and Epstein 2010). There are notable parallels here, of course, with heterogeneous network concepts such as actor-network theory and *agencement* (see the introduction to Part I), which embrace the idea that non-humans (such as trees, forests) have agency. Trees are living things and can catch fire, succumb to disease, be hollow (and therefore store much less carbon), hide beneath clouds (making satellite measurements difficult), be blown over, have more/less roots than supposed and so on. Hence, a socio-technical perspective on the case of forest carbon MRV suggests forests are perhaps not amenable to being enrolled or translated into markets and economies in the way that other artefacts might be, and that is why measuring carbon in forests to compliance-grade standard for REDD+ has proved to be so challenging. Clearly forests – comprised of trees and a multitude of other living things – are not durable and stable in the same way that, for example, buildings, power stations or rail tracks are. Indeed, much recent forest science has highlighted the fragility of tropical forest ecosystems (see Malhi *et al.* 2008). A sociotechnical perspective also helps explain why so much attention and resources have been invested in forest carbon MRV. This is because the rewards are so great: managing the inherent risks of forests through a comprehensive MRV framework would enable forest carbon to become commensurate with other types of carbon already relatively stabilised within international carbon markets (see Chapter 2).

Summary and conclusions

Chapter 6 has introduced and discussed another example of the making of low carbon economies – namely, managing and measuring the carbon in tropical forests, a crucial step in bringing new forest carbon markets into being. The thrust of REDD+ policy is about making forests governable through markets, bringing them into the realm of international carbon markets through numerous acts of commensuration. The specific appeal of remote sensing can be better understood by conceptualising forest carbon MRV as a type of governmentality; a governmentality enabled through obtaining forest carbon data and applying particular types of expert knowledge. A focus on commensuration is a useful entry point to understanding how new markets and economies are made, involving a complex mix of expertise, technologies, policies and politics. Acts of commensuration necessarily have to leave out data and ideas that do not fit: the objective of standardisation is creating simplified categories and rules, and in the process discarding things, ideas, knowledge. So, whilst the multiple standards emerging in forest carbon MRV provide valuable stability, capturing the uncertainty and dissent amongst the multiple expert communities on forests is also important if a long-term, workable solution to forest carbon MRV is to be developed. The forest carbon case study in this chapter provides a good illustration of both the challenges and opportunities in making low carbon markets functional in an area of operation characterised by fast-moving science. Governmentality perspectives on the role of knowledge, and how and in what circumstances certain types of knowledge becomes widely accepted (and therefore powerful), are key to understanding how forest carbon market MRV has evolved to date, centred on remote sensing.

Forest ecosystems are inherently dynamic, diverse and often fragile. Ideas about acts of commensuration, especially their sociotechnical characteristics (see the introduction to Part III), usefully draw our attention to the 'stuff' of climate policy – whether it be forests or heat meters (see Chapter 3) – reminding us that artefacts and living things are not necessarily pliable or reliable and therefore perhaps not so amenable to being brought into, and 'behaving' within, markets. A strength of sociotechnical approaches is in explaining some of the inconsistencies and difficulties inherent to measuring forest carbon, with a focus on the trees and forests themselves: without close attention to the substance of what is being measured and monitored, there is a danger that new low carbon market objectives may be overly ambitious. New low carbon actors such as forests have the capability to significantly disrupt markets. Acts of commensuration such as forest carbon MRV attempt to stabilise and harmonise such behaviours and thereby do vital 'behind the scenes' work in the making of new low carbon markets and economies. However, an excessive focus on forest carbon MRV has necessarily limited the scope of forest carbon markets: ideas and institutions have been somewhat marginalised, such as more informal and qualitative community-based forestry techniques and organisations, because these techniques and types of expertise cannot so easily be converted into acts of commensuration.

Notes

1 The term 'REDD' has over time evolved to 'REDD+', with the '+' signifying activities related to the sustainable management and enhancement of forest carbon stocks (the text after the semi-colon in the Bali Action Plan – namely, "... the role of conservation, sustainable management of forests and enhancement of forest carbon stocks in developing countries" (UNFCCC 2007: Decision 1/CP.13, para 1b iii).
2 An expert science body established by the United Nations in 1988 to provide advice on climate change to governments.
3 Including the Miombo International Conference, Edinburgh, June 2008; the Commonwealth International Forestry Conference, Edinburgh, July 2010; the UK Government Earth Observation Workshop, Reading, June 2010; and Forest Day 6 at the UNFCCC annual meeting (COP/MOP), Doha, Qatar, November 2012.

Bibliography

Backstrand, K. and E. Lovbrand (2006). "Planting trees to mitigate climate change: contested discourses of ecological modernization, green governmentality and civic environmentalism". *Global Environmental Politics* **6**(1): 50–75.
Baker, J. D., K. Andrasko, M. Bartlell, R. deFries, M. Herold, D. Murdiyarso, D. Pandey, D. Singh and C. Souza Jnr (2010). *MRV for REDD+ for Classes of Countries: Requirements, Readiness and Capacity Gaps*. Washington, World Bank, Forest Carbon Partnership Facility.
Barry, A. (2006). "Technological zones". *European Journal of Social Theory* **9**(2): 239–253.
Boyd, E., M. Gutierrez and M. Chang (2007). "Small-scale forest carbon projects: adapting CDM to low-income communities". *Global Environmental Change* **17**: 250–269.
Bryant, R. L. (2002). "Non-governmental organisations and governmentality: 'consuming' biodiversity and indigenous people in the Phillippines". *Political Studies* **50**: 268–292.
Bulkeley, H. and M. Betsill (2003). *Cities and Climate Change: Urban Sustainability and Global Environmental Governance*. London and New York, NY, Routledge.
Clinton Climate Initiative (2012). "Preserving forests: a conversation with Dr. D. James

Baker". Online at: www.clintonfoundation.org/what-we-do/clinton-climate-initiative/i/preserving-forests-a-conversation-with-dr-d-james-baker (accessed 23 January 2012).

Dean, M. (1999). *Governmentality: Power and Rule in Modern Society*. London, Sage Publications Ltd.

Dryzek, J. (1997). *The Politics of the Earth: Environmental Discourses*. Oxford, Oxford University Press.

Edwards, P. N. (2004). "'A vast machine': standards as social technology". *Science* **304**(5672): 827–828.

Eggleston, H. S., L. Buendia, K. Miwa, T. A. Ngara and K. Tanabe (2006). "IPCC guidelines for national greenhouse gas inventories – Volume 4 Agriculture, Forestry and Other Land Use". Kanagawa, IPCC, National Greenhouse Gas Inventories Programme, Institute for Global Environmental Strategies.

Fogel, C. (2005). "Biotic carbon sequestration and the Kyoto Protocol: the construction of global knowledge by the Intergovernmental Panel on Climate Change". *International Environmental Agreements: Politics, Law and Economics* **5**(2): 191–210.

Foucault, M. (1991). "Governmentality". *The Foucault Effect: Studies in Governmentality*. G. Burchell, C. Gordon and P. Miller (eds). London, Harvester Wheatsheaf: 87–104.

Global Canopy Programme (2009). "The little REDD+ book – an updated guide to governmental and non-governmental proposals for reducing emissions from deforestation and degradation". Oxford, Global Canopy Programme.

Global Carbon Project (2013). "Global carbon budget 2013 – summary". Online at: www.globalcarbonproject.org/carbonbudget/13/hl-compact.htm (accessed 11 February 2014).

Global Observation of Forest and Land Cover Dynamics (GOFC-GOLD) (2009). "A sourcebook of methods and procedures for monitoring and reporting anthropogenic greenhouse gas emissions and removals caused by deforestation, gains and losses of carbon stocks in forests, remaining forests and reforestation. Report version COP15-1". Geneva, Global Observation of Forest and Land Cover Dynamics (GOFC-GOLD).

Global Observation of Forest and Land Cover Dynamics (GOFC-GOLD) (2012). "GOFC-GOLD: Global Observation of Forest and Land Cover Dynamics – home page". Online at: www.fao.org/gtos/gofc-gold/ (accessed 30 July 2012).

Grainger, A. (2009). "Towards a new global forest science". *International Forestry Review* **11**(1): 126–133.

Group on Earth Observations (GEO) Forest Carbon Tracking (2010). "Report on the concept phase for developing a global forest monitoring network". Geneva, Group on Earth Observations (GEO) Forest Carbon Tracking Task.

Hajer, M. A. (1995). *The Politics of Environmental Discourse: Ecological Modernisation and the Policy Process*. Oxford, Clarendon Press.

Herold, M. (2009). "An assessment of national forest monitoring capabilities in tropical non-Annex I countries: recommendations for capacity building". London and Oslo, The Prince's Rainforests Project and the Government of Norway.

Laws, D. and M. Rein (2003). "Reframing practice". *Deliberative Policy Analysis: Understanding Governance in the Network Society*. M. A. Hajer and H. Wagenaar (eds). Cambridge, Cambridge University Press: 172–206.

Li, T. M. (2007). *The Will to Improve: Governmentality, Development, and the Practice of Politics*. Durham, NC, and London, Duke University Press.

Lovbrand, E. and J. Stripple (2010). "Governing the climate from space: monitoring, reporting and verification as ordering practice". Paper presented at the 2010 International Sociological Association Annual Convention, New Orleans, 17–20 February.

Lovell, H. (2014, forthcoming). "The multiple communities of low carbon transition: an assessment of communities involved in forest carbon measurement". *Local Environment*. Online at: www.tandfonline.com/doi/abs/10.1080/13549839.2014.905515?journalCode=cloe20#.U_8RgBDpxuJ (accessed 27 August 2014).

Lovell, H. and D. MacKenzie (2014). "Allometric equations and timber markets: an important forerunner of REDD+?" *The Politics of Carbon Markets*. B. Stephan and R. Lane (eds). London, Routledge: 69–90.

Luke, T. (1995). "Environmentality: geo-power and eco-knowledge in the discourses of contemporary environmentalism". *Cultural Critique* **31**, The Politics of Systems and Environments, Part II (Autumn): 57–81.

Luke, T. (1999). "Environmentality as green governmentality". *Discourses of the Environment*. E. Darier (ed.). Oxford, Blackwell Publishers: 121–151.

Malhi, Y., J. T. Roberts, R. A. Betts, T. J. Killeen, W. Li and C. A. Nobre (2008). "Climate change, deforestation, and the fate of the Amazon". *Science* **319**(5860): 169–172.

Owens, S. E. and R. Cowell (2002). *Land and Limits: Interpreting Sustainability in the Planning Process*. London, Routledge.

Penman, J., M. Gytarsky, T. Hiraishi, T. Krug, D. Kruger, R. Pipatti, L. Buendia, K. Miwa, T. Ngara, K. Tanabe and F. Wagner (2003). "IPCC Good Practice Guidance for land use, land-use change and forestry". Institute for Global Environmental Strategies, Kanagawa, National Greenhouse Gas Inventories Programme.

Picard, N., L. Saint-André and M. Henry (2012). "Manual for building tree volume and biomass allometric equations: from field measurement to prediction". Rome, Food and Agricultural Organization of the United Nations, Montpellier, Centre de Coopération Internationale en Recherche Agronomique pour le Développement.

Rein, M. and D. Schon (1993). "Reframing policy discourse". *The Argumentative Turn in Policy Analysis and Planning*. F. Fischer and J. Forester (eds). London, UCL Press Ltd.: 145–166.

Stephan, B. (2012). "Bringing discourse to the market: commodifying avoided deforestation". *Environmental Politics* **21**(4): 621–639.

Tanaka, K. and L. Busch (2003). "Standardization as a means for globalizing a commodity: the case of rapeseed in China". *Rural Sociology* **68**(1): 25–45.

Timmermans, S. and S. Epstein (2010). "A world of standards but not a standard world: toward a sociology of standards and standardization". *Annual Review of Sociology* **36**: 69–89.

United Nations Programme on Reducing Emissions from Deforestation and Forest Degradation (UN REDD Programme) (2008). "Role of satellite remote sensing in REDD: issues paper". Rome, FAO, UNDP, UNEP.

United Nations Framework Convention on Climate Change (UNFCCC) (2007). "Report of the Conference of the Parties on its thirteenth session, held in Bali from 3 to 15 December 2007". Bonn, United Nations Framework Convention on Climate Change (UNFCCC).

United Nations Framework Convention on Climate Change (UNFCCC) (2009). "Copenhagen Accord: Decision 4/CP.15 Methodological guidance for activities relating to Reducing Emissions from Deforestation and Forest Degradation and the role of conservation, sustainable management of forests and enhancement of forest carbon stocks in developing countries". Copenhagen, UNFCCC.

Watson, C. (2009). "Forest carbon accounting: overview and principles". Addis Ababa, Ethiopia, UNDP-UNEP CDM Capacity Development Project for Eastern and Southern Africa.

Williams, M. (2008). "Quantifying and monitoring carbon stocks in tropical woodlands". Paper presented at the Carbon and Communities in Tropical Woodlands: An International Interdisciplinary Conference, University of Edinburgh, Edinburgh, 16–18 June.

7

THE TREATMENT OF CARBON IN FINANCIAL ACCOUNTS

Key findings

- Financial accounting standards are largely invisible to those working on carbon markets, but nonetheless provide the calculative bedrock underpinning these markets.
- The study of commensuration within low carbon economies is not just about the development of new types of commensuration (as in Chapter 6), but also about how existing forms of measurement, calculation, classification and standards are changing and evolving in response to climate change.
- Carbon is being understood, framed and classified with reference to other more familiar things known to financial accountants, such as government grants, taxes and leases.
- In the absence of International Accounting Standards for carbon, a number of 'coping strategies' are being used, including seeking authority elsewhere (i.e. outwith the International Accounting Standards Board (IASB)), comparison of carbon to other commodities, lobbying the IASB for standardisation and reducing disclosure.

Introduction

In this chapter it is demonstrated how a fuller understanding of both the potential and the weaknesses of carbon markets requires investigation of their implications for other disciplines and professional activities: interdisciplinary work on financial accounting is crucial in this respect and to date has been somewhat overlooked by researchers in fields other than accounting (for exceptions, see Hatherly *et al.* 2008; Lohmann 2009; MacKenzie 2009; Ascui and Lovell 2011; Lovell and MacKenzie 2011). Here, attempts by accountants to classify carbon and set accounting rules for greenhouse gas emission allowances (hereafter 'emission allowances') are explored. The empirical analysis, which draws on findings from in-depth interviews with accountants, a survey of European companies and analysis of the board meetings and 2011 Agenda Consultation of the International Accounting Standards Board (IASB), is of interest both in understanding the detail of how markets, such as for carbon, come into being, as well as providing new insights into the role of standards within markets and the policy process. It is demonstrated that emission allowances have been a difficult thing for accountants to make a judgement on – an 'incommensurable' – and accountants have sought to make them commensurable using a range of techniques and practices, including frequent comparison with other, more familiar accounting items (such as government grants, taxes and leases) and, in the absence of international standards, seeking guidance and authority elsewhere (from national accounting organisations, industry bodies and auditors).

Whilst the IASB (the standard setting body of the International Financial Reporting Standards Foundation (IFRS)) might seem at first glance to be a rather mundane, unlikely location for conducting climate change research, its innocuous boardroom in central London belies the importance that decisions made here have for the operation of carbon markets (and, of course, other markets and business operations) worldwide. The IASB comprises fifteen full-time members (fourteen men; one woman), all highly regarded in the financial and business worlds, whose job is to reach impartial decisions on a wide range of accounting issues, meeting every month for five days. In the world of accounting, IASB decisions are carefully monitored, and the IASB is paid close attention to: one can register to attend the meetings in person as an observer or listen to the live webcast or summary podcasts. For instance, in the May 2012 board meeting, amongst other things, the future accounting rules for insurance contracts, leases and macro hedge accounting were all decided upon (IFRS 2012), making this particular London boardroom an important, if slightly unusual, site of empirical research into standard setting.

This chapter draws on in-depth interviews (17) with accountants active in carbon financial accounting in a range of different types of organisation, including: large accountancy firms, standard setters (the IASB and its US counterpart – the Financial Accounting Standards Board, FASB) and at large European companies involved in emissions trading. The interview material has been supplemented by analysis of IASB board meeting minutes, the response letters (248) to an IASB 2011 Agenda Consultation, as well as a financial report survey. The financial report survey, undertaken in 2010, involved gathering data on the accounting disclosures made in the financial statements of the largest greenhouse gas emitting companies in Europe (26), in order to ascertain the systems of classification, measurement and reporting followed by the companies (see Lovell *et al.* 2010 for further detail). I also was a member (2009–2012) of a Technical Working Group of an international carbon accounting organisation, and this work, including attending monthly committee meetings and providing expert input to reports and strategy, has informed and provided valuable empirical context.

The chapter is structured as follows. First, ideas from relevant scholarship on commensuration, especially those pertaining to standards, are interweaved with an analysis of international carbon financial accounting standards since the advent of the European Emissions Trading Scheme (EU ETS) in 2005. Through a necessarily partial overview of the history of accounting standard setting for emissions trading, it is hoped that a flavour of the technical complexity of the financial accounting ambiguities relating to emission allowances is given. Emission allowance accounting is an area of standard setting that remains in a state of flux, despite repeated attempts to set rules and guidelines. Second, in the main empirical section of the chapter, precisely how accountants are seeking to define and understand emission allowances is critically assessed, focusing on the coping strategies they are using in the prolonged absence of an international standard. Third, in conclusion, the discussion centres on the value of focusing on standards and other acts of commensuration to research the making of low carbon economies, because it helps reveal the sociotechnical messiness of adapting existing standards and making new ones. Standards are at work in markets and economies, government, business and elsewhere, and understanding their role is important.

Standards, financial accounting and emission allowances

Scholarship on commensuration is a natural starting point for thinking about financial accounting simply because, first and foremost, accounting is an area of activity structured by a significant and complex array of principles, rules and standards. Theories about standards and other acts of commensuration come from diverse disciplinary origins: from education, accountancy and business studies, science and technology studies and sociology (Cronon 1991; Abbatte 1999; Bowker and Star 2000; Tanaka and Busch 2003; Edwards 2004; Kolk *et al.* 2008; Higgins and Larner 2010; Timmermans and Epstein 2010). In this section, I recap and build on elements of this scholarship, as outlined in the introduction to Part III, with particular relevance to this case study, using these ideas to introduce the topic of financial accounting.

Financial accounting clearly fits into the category of 'big', stable, international standards: a set of standards that have temporal depth and geographical reach, underpinning economic activity worldwide. However, in line with arguments about the hidden day-to-day work of standards, the process of setting financial accounting standards is not a topic that has met with great interest by the wider social science community (albeit with some notable exceptions; see Hopwood and Miller 1994; MacKenzie 2008). In common with other types of standards, financial accounting standards use highly expert technical language, largely impenetrable to non-accountants. The IASB, based in London, UK, was established in 1973 (then called the International Accounting Standards Committee, renamed in 2001) and its overall goal is "to develop a single set of high quality, understandable, enforceable and globally accepted financial reporting standards based upon clearly articulated principles" (IASB 2012b). It does this via a series of over fifty interrelated International Financial Reporting Standards (IFRSs), used by over 100 countries worldwide. The IASB, however, is not the only international financial standard setter – its counterpart in the United States is called the Financial Accounting Standards Board (FASB), and it also has an extensive set of standards, called 'Generally Accepted Accounting Principles' (GAAP), which are used across the United States and beyond.

Standardisation of new 'things' is always done with reference to what has already been standardised – to the existing body of standards – but at the same time existing standards and classifications are always subject to revision: there is a constant tension between continuity and change. For emission allowances entering the world of financial accounting, we have a case of a new thing needing to be assessed and made to fit with an existing, well-embedded and extensive set of standards, and it is at this stage in the process of setting standards for emission allowances that difficulties have emerged. In the run-up to the advent of the European Emissions Trading Scheme (the EU ETS) in 2005, accounting guidance was issued by the IASB via its International Financial Reporting Interpretations Committee (IFRIC). "IFRIC Interpretation 3: Emission Rights" ("IFRIC-3") was published in December 2004. IFRIC-3 classified emission allowances as *intangible assets* (whether allocated for free by governments or purchased), therefore falling under International Accounting Standard (IAS) 38. However, accounting for emission allowances required the use of more than one standard, as there was the complicating issue of governments (in the early stages of the EU ETS – Phases 1 and 2) giving allowances out for free to companies. This politically driven free allocation needed to be interpreted with care by financial accountants. IFRIC-3 ruled that allowances distributed for free should be measured initially at their fair value (i.e. their market price, which at the time of writing stands at around €5 per tonne of carbon dioxide, but historically has been as high as €20 per tonne), and that the difference between the amount paid (zero, as issued for free) and fair value should be classified as a government grant and therefore accounted for under IAS 20 ("Government grants and disclosure of government assistance"). In terms of liabilities, it was judged that a liability should be recognised as emissions are made, and that it should be a 'provision', and thus to be accounted for following IAS 37 ("Provisions, contingent liabilities and contingent assets"), and measured at fair value.

Whilst much of the literature on standards concentrates on the reach and powerful day-to-day operation of existing large bodies of standards (health, weather, etc.), scholars have suggested that most can be learnt from cases of things that do not fit with these existing standards (what Bowker and Star (2000) call 'incommensurables': new markets/commodities/things that pose difficulties in being measured, classified and standardised) (see the introduction to Part III). The case of emission allowances within financial accounting standards is a clear example of such an 'incommensurable'. As is explained in more detail below, the use of multiple accounting standards for emission allowances (IAS 38, 20 and 37), as proposed by IFRIC-3, quickly ran into problems and has mostly not been followed by companies reporting on their emission allowances (Lovell *et al.* 2013). The detail of the accounting interpretation in IFRIC-3 is important here, in so far as it indicates that for emission allowances the pathway from classification to standardisation has not been a smooth one. In part, as noted, this is because of the political decision taken in Europe for governments initially to not make companies pay for emission allowances. But it is also because, for some stakeholders, carbon financial accounting raises profound questions about the definitions and principles embodied within the IFRSs (i.e. the overarching financial accounting standards), a point returned to later.

Standards are devised by technical experts in the field, whether it be engineers, farmers, plumbers or doctors. These experts have profound knowledge in the particular field that the standard relates to, but not typically beyond its boundaries. This narrow framing of standard setting expertise is the source of some of the problems in the case of carbon financial accounting: carbon markets are not a topic familiar to financial accounting standard setters. It is possible to view standards as a type of market or calculative device – "the material and discursive assemblages that intervene in the construction of markets" (Muniesa *et al.* 2007: 2; see the introduction to Part I) – essential to the practical work of bringing a new market like carbon into being. Standardisation is thus a way of 'pacifying goods' so they are fit for markets, so they remain stable. However, if you do not have a well-designed comprehensive market system (including standards) that encompasses all component parts – i.e. that reaches across different bodies of expertise – then this leads to 'overflowing' (Callon 1998; Callon and Muniesa 2007). These ideas suggest, therefore, that part of the reason for accountants not agreeing on a financial accounting standard is a failure of carbon markets to be designed comprehensively (i.e. including financial accounting). Indeed, in the numerous histories of the formation of the EU ETS (see, for example, Grubb 2002; Bailey and Wilson 2009), it is clear that its conception was rather rushed and subject to much political bargaining – two characteristics that suggest a lack of attention to the long-term day-to-day operation of the EU ETS and, in particular, the types of expertise required therein.

The case of carbon financial accounting demonstrates the politics of standard setting. This is most clearly illustrated via the degree of protest and uproar (within the financial accounting community at least) in the aftermath of the IASB releasing IFRIC-3 (see Bebbington and Larrinaga-Gonzalez 2008; Cook 2009;

MacKenzie 2009 for a more in-depth discussion). The European Financial Advisory Group (EFRAG) issued particularly negative advice, which carried considerable weight, given that the carbon market that prompted IFRIC-3 – the EU ETS – was based in Europe (Bebbington and Larrinaga-Gonzalez 2008). EFRAG's objections were focused on accounting mismatches in the IFRIC-3 recommendations, with concern that they would lead to an artificial volatility of company results. In particular, there was negative reaction amongst major EU ETS participants (utilities and large industry emitters, those institutions represented by EFRAG) on a number of grounds, including where to account for carbon (with IFRIC-3 recommending some gains and losses to be reported in the income statement and some in the equity statement, what is known as a 'mixed presentation model') and how to balance assets and liabilities (with IFRIC-3 recommending some emission allowances to be measured at cost – i.e. the price paid for them – and others at fair value – at the current market value – known as a 'mixed measurement model' (Cook 2009; MacKenzie 2009)). In short, the method of classifying and standardising carbon proposed by the IASB in IFRIC-3 was judged in practice to be unworkable by those companies who would have to actually do the accounting: the multiple identities of carbon were not stabilised within IFRIC-3; there were too many 'overflows' from this particular framing (Callon 1998), and IFRIC-3 was withdrawn in mid-2005, leaving a gap in International Accounting Standards for emission allowances that still remains.

It was not until 2008 that any work on emissions trading was restarted by the IASB, this time in the form of a joint project with its US counterpart, the Financial Accounting Standards Board (FASB). The IASB-FASB joint Emissions Trading project had a somewhat broader remit than IFRIC-3, in part reflecting the involvement of both organisations, plus a dramatic rise in the number of new and planned carbon markets worldwide at that time (Point Carbon 2007; IASB 2008). The project was therefore not just about the EU ETS, but aimed to address the accounting of all tradable emissions rights and obligations arising under any emission trading scheme worldwide. Some progress was made with this project before it was 'paused' in 2010, because of a fundamental IASB review of their project load and objectives. The 2010 IASB Strategy Review resulted in the cessation of most of its standard setting projects (in other words, the Emissions Trading project was not singled out for special attention; the Strategy Review was prompted by the 2008 financial crisis and the role of financial accounting therein). However, the Emissions Trading project has (as of May 2012) been identified as a project where further research is required (IASB 2012a: 13), giving some optimism to those urging the IASB to develop an international standard.

Drawing on Foucault's notion of governmentality (see the introductions to Part I and Part III), connections between 'rationalities' (expertise, ways of thinking and discourse) and 'techniques' (standards and other acts of commensuration) are further illuminated. Such an approach posits that carbon financial accounting governance and practice can only be understood through examining the discursive rationale for the IASB in conjunction with its technical standards, as well as the framing and day-to-day operation of carbon markets. Applied to the case of carbon financial accounting, the practical, technical difficulties accountants are experiencing in classifying and managing emission allowances are thus linked to underlying ambiguities and tensions in the policy discourse and operation of carbon markets and financial accounting standards. Using a governmentality approach, financial accounting standards are transformed from something mundane and mostly hidden into a set of technologies with the potential ability to effect change in the discourse, framing and even the whole notion of carbon markets.

With some detail of the history of emission allowance accounting already covered, this chapter now turns to consider in more depth how and why emission allowances have proved so difficult to fit into existing accountancy rules and standards, and what the response of accountants has been to a prolonged absence of standards.

Coping strategies: how accountants are responding to an absence of international carbon accounting standards

As noted, issues of financial accounting market practice are invisible to the public (and probably many policymakers), but nonetheless have a significant influence on the fungibility of carbon and the overall effectiveness of carbon markets. In this empirical section, I concentrate on how accountants are seeking to resolve ambiguity about how to measure, classify and disclose emission allowances. A number of coping strategies are at work, which I discuss in turn, including: seeking authority elsewhere (i.e. outwith the IASB); comparison of carbon to other commodities; lobbying the IASB for standardisation; and reducing disclosure. The analysis provides useful empirical insights into what happens in the absence of international standards and other acts of commensuration in an area where one might otherwise expect them, such as financial accounting. This is a topic not well-covered by existing scholarship on standards, which tends to focus on existing stabilised standards and the emergence and subsequent integration of new 'things', and not typically situations of prolonged non-integration (see the introduction to Part III).

Seeking authority elsewhere

There is evidence that practitioner accountants (i.e. those 'at the coalface', doing the day-to-day work of producing company accounts) are looking elsewhere – beyond the IASB – for authority and reassurance regarding their choice of emission allowance accounting treatment and disclosure. For instance, the global accountancy firm Deloitte, in their regular newsletter on IASB decisions *IASplus*, in early 2012 made the observation that national standard setters are 'moving forward on their own' in the continuing absence of international standards for emission allowance accounting (Deloitte 2012). Further, the French national accounting standard setter – Autorité des Normes Comptables (ANC) – published a guidance document called "Proposals for accounting of greenhouse gas emission rights reflecting companies' business models" in October 2012 (ANC 2012), setting out in detail its recommended accounting treatment for emission allowances. It advises flexibility for companies to enable them to vary their accounting according to whether their emission allowances are held for trading or to comply with regulation (a so-called 'economic' or 'activity-based' approach that the IASB does not support). Further, the Australian national accounting body – the AASB – similarly released a Staff Discussion Paper in 2012, setting out detailed accounting guidance for Australian companies and organisations preparing for new carbon market legislation in Australia (AASB 2012).

In addition to the work of national accounting organisations, industry groups have also developed guidance in sectors where carbon markets have a significant influence – namely, the electricity sector and energy utilities. Most notable here in the European context is the accounting guidance produced by the International Energy Accounting Forum (IEAF) – a group of energy industry accountants based in Europe, who have produced, and follow, their own set of emission allowance accounting rules (IEAF 2010). Similarly to the French national accounting body guidance, the IEAF recommends organisations account for emission allowances differently, according to whether the allowances are held for production (own use) purposes or trading.

The role of auditors has also been important in providing advice and guidance in the absence of international standards, as an interviewee explains:

> You will probably find if they [companies] have the same auditor, they will be doing it [the carbon financial accounting] the same way . . . most companies will have consulted with their auditors and said "Hey, we have got this new commodity, how the heck do we account for this?"
>
> (Interview, Director – Energy and Utilities Group at a large international accountancy firm, November 2008)

Auditors have, in keeping with national and sectoral standards, generally advocated an 'activity-based' model of accounting. For example, the accountancy firm KPMG formalised this approach in its carbon accounting guidance (KPMG 2008), advising accountants to follow different accounting rules depending on the type of organisation they are located in (and hence the type of activity being undertaken). The guidance report classifies organisations according to their dominant activity, as: emitters, creators/green energy, trader/aggregators and investors/consultants. However, significantly, the 'activity-based' solution to emission allowance accounting runs contrary to the opinion of the IASB and lies at the heart of difficulties in resolving the situation. The IASB has stated that it disagrees with any approach (such as the 'activity-based' model) that differentiates between the production and trading of emission allowances, as the manager at the IASB with responsibility for the Emissions Trading project (2008–2010) explained in an interview:

> I think a lot of [IASB] Board Members would say well, if you have one thing [an emission allowance], to look at it in different ways doesn't make a lot of sense. Because if you have two allowances in an activity-based model you account for one in that direction and the other in that direction. But why are they so different? You can interchange them [the allowances] and well, you can even change your mind. Just because you said at the outset, well I'm going to trade that instrument, perhaps you might use it later on in your production process.
>
> (Interview, IASB Manager, November 2008)

The rationale for the IASB's position is that even though emission allowances are used in different ways, this should not be the basis for International Accounting Standards, because they are still the same thing. In other words, even if emission allowances are classified temporarily by 'users' (companies, financial exchanges, traders) for distinctive purposes, they should be treated by accountants consistently in a single, uniform way. This tricky combination of the potential of emission allowances for multiple uses and their fungibility lies at the heart of the struggle by accountancy standard setters to agree on a standard. At the root of the problem of stabilising the identity and definition of emission allowances is this question of how to manage their multiple uses: individuals and organisations legitimately use emission allowances in different ways – to comply with regulation, voluntarily offset their emissions, trade and make profits. Indeed, carbon markets such as the EU ETS were designed precisely with this flexibility in mind; the policy discourse of markets versus state-based 'command and control' regulations hinged on giving choice and flexibility to corporations via markets (Kolk *et al.* 2008; Bailey and Maresh 2009). In practice, however, we observe friction at the intersection of financial accounting and carbon markets, where different understandings of, and expectations about, the purpose and identity of emission allowances jostle. The conceptual, principles-based IASB approach sits uneasily with carbon financial accounting 'in the wild' (see Callon 2007: 338), where practical decisions have had to be made in the prolonged absence (since 2005) of an accounting standard. But, contrary to suggestions about the problem of 'overflows' to successful market operation (Callon 1998, 2009), the diversity of standards emerging on emission allowance accounting could be viewed as a strength, rather than a problem: expert communities of practice (national/sectoral) are developing standards that fit their needs, allowing at least for some degree of co-ordination, consultation and learning.

Comparison of carbon to other commodities

A second response of organisations to a lack of international standards in this area has been to seek authority for their accounting approach through a different route – namely, through comparing emission allowances to other commodities or accounting entities where accounting rules have been decided, such as taxes and leases. For example, in a 2009 IASB board meeting discussion paper (December; "Agenda paper 18": IASB 2009), a number of points of comparison were made between emissions trading and other accounting issues, in order to help guide board members in reaching a decision:

> The issue of a government having discretion to subsequently change access to a restricted resource (ie allocation) is not unique to emissions trading schemes. Governments, for example, grant access to restricted resources by, *for example, transferring or allocating airport landing rights, licenses to operate radio or television stations, import licenses or quotas . . . airport landing rights are often conditional on the airline continuing to make use of the landing rights. This is not different to an entity's right to future instalments in an emissions trading scheme that is conditional on an entity continuing its emitting operations.*
>
> (IASB 2009: para 31 [emphasis added])

In a number of other locations and discussions beyond the IASB, the comparison of emission allowances to other commodities similarly takes place. In the 2012 French ANC accounting guidance, for example, it states that:

> [Emission allowances] fall outside the definition of intangible assets as, although they are without physical substance, they do not have the features common to known intangible items (such as *fishing quotas, software, taxi licences, reproduction rights*), especially in the context of emission rights requiring payment.
>
> (ANC 2012: 7 [emphasis added])

So, whilst it might seem somewhat unusual at first glance (for non-accountancy experts) to draw links between emission allowances and fishing quotas or reproduction rights, to accountants this is a logical and understandable strategy and, indeed, in keeping with findings from other studies of standards, where attempts to draw analogies are a key part of the processes of labelling and classification (Espeland and Stevens 1998; Bowker and Star 2000). It is, in essence, a practice-based response to the ongoing ambiguity, with corporate accountants and practitioners 'muddling through' to derive a method of accounting that appears to work for them and allowing emission allowances to be 'made passive', even if it sits uneasily with the IASB's principles-based approach to accounting standard setting.

Lobbying the IASB

A third response of accountants and other interested stakeholders to the absence of IASB standards has been to lobby the IASB to put rules in place. Whilst lobbying activity has been somewhat muted (perhaps understandable in the wider context of the global financial crisis in the period 2008–2012 and the associated upheaval in financial accounting), there is evidence of reasonably strong support for the IASB to take action and restart its Emissions Trading project. This comes in the form of responses to an IASB Agenda Consultation exercise in 2011. Out of the 248 consultation response letters received by the IASB, thirty-eight of these (12 per cent) asked (unprompted) for the IASB to restart its Emissions Trading Schemes (ETS) project and issue guidance on emission allowance accounting. However, balanced against this, 9 per cent of respondents (23) explicitly said that this project was a low priority and should not be restarted. Table 7.1 demonstrates this diverse range of views for and against the IASB restarting work on emission allowances using illustrative extracts from IASB 2011 Agenda Consultation response letters.

The profound level of disagreement shown in Table 7.1 regarding to what extent an absence of standards is a problem or not is striking. Also notable is that for those against the IASB restarting the ETS project, the majority of concerns relate to the potential for emission allowance accounting to unravel the existing set of international financial accounting standards (IFRSs), with comments such as "...it highlights that existing IFRSs are deficient" (AASB, Consultation Letter 237) and that "...developing a satisfactory, principle-based IFRS [for emission allowances] may not be possible without considering the underlying IFRSs and conceptual framework" (Grant Thornton, Consultation Letter 77). The official IASB response to the Agenda Consultation identified the ETS as a "priority research project" (IASB 2012a: 11) (one of only nine such projects), signalling at least some support for developing a standard.

Reducing disclosure

A fourth and final response of accountants to the lack of international guidance has been, simply, to provide less information in the public domain about their emission allowances. In the absence of International Accounting Standards, there is no formal requirement for organisations to provide emission allowance data in their financial reports, and so many companies consequently have very low levels of disclosure. A survey of carbon financial accounting practices in 2010 found that 69 per cent of companies surveyed did not disclose any information about depreciation of emission allowances, and 23 per cent provided no information about their emission allowance liabilities (Lovell *et al.* 2010). One accountant described the process they have been through at his company of gradually reducing emission allowance disclosure:

> At first we wanted to be very transparent, we wanted to disclose everything, yes?... But because we started to have a lot of problems with our auditors we decreased the disclosure ... we haven't changed the accounting scheme, which could have material impact on numbers, but we changed the disclosure ... we decreased the disclosure very, very much and now it is, you know, it's good.
>
> (Interview, deputy head of accounting at a large European energy utility, May 2010)

Worryingly, this reduction in disclosure is despite these emission allowances being material (i.e. financially significant) to company accounts (see Lovell *et al.* 2013), meaning that investors and other stakeholders are not being provided with market-relevant information. This curbs the transparency and effectiveness of the EU ETS and runs contrary to the fundamental objective of carbon markets to act as an efficient and transparent means of reducing greenhouse gas emissions.

Summary and conclusions

Conceptualising carbon accounting potentially cuts across a number of different theories and bodies of research. There is a range of relevant literature to draw upon that offers useful insights into how accountants might be deciding upon what kind of an entity emission allowances are, how to classify emission allowances and what happens in the absence of international standards. An obvious starting point in thinking about financial accountancy – an area of professional expertise somewhat renowned for rules, measurement and classification – is current scholarship on commensuration. This modest literature provides much in the way of detailed empirical case studies of standard setting and standardisation, ranging from global weather measurement to electrical voltages and international disease classification (see the introduction to Part III). It touches on key themes identified in the empirical case: classification, stability and expertise.

Table 7.1 Extracts of the IASB 2011 Agenda Consultation response letters related to emissions trading

Author of IASB 2011 Agenda Consultation Letter (CL) (inc. IASB CL reference)	*IASB should prioritise/restart the Emission Trading Schemes (ETS) project*	*Author of IASB 2011 Agenda Consultation Letter (CL) (inc. IASB CL reference)*	*IASB should NOT prioritise/restart Emission Trading Schemes (ETS) project*
International Public Sector Accounting Standards Board (#CL29)	"[W]e support the IASB reactivating this [ETS] project, given the number of jurisdictions in which ETSs have been implemented and the number of entities affected by such schemes . . ."	Grant Thornton (#CL77)	"We recognise that these schemes have grown in importance and that applying existing IFRSs is challenging. However, we feel that developing a satisfactory, principle-based IFRS may not be possible without considering the underlying IFRSs and conceptual framework."
Deloitte (#CL43)	"Given the pervasiveness of the issue and the current lack of guidance in IFRSs on dealing with such schemes, there is a considerable risk of divergent practices emerging and of national or regional regulators introducing their own requirements to fill the gap in IFRSs, thus diluting the status of IFRSs as a comprehensive set of financial reporting standards without regional variation. Consequently, we think a narrow-scope project that would achieve consistency in the short-term is necessary. In the longer term, we think that the intangibles project [IAS 38] should include ETS within its scope."	Royal Dutch Shell (#CL165)	"As the Board already knows from its experience with IFRIC 3 and its subsequent research, the accounting for ETS . . . raises a number of complex conceptual issues and any proposals (for example, by amending IAS 38 to allow emissions allowances to be carried at fair value with changes therein recognised in profit or loss rather than other comprehensive income) would have consequential effects that would most likely cause the project to expand beyond its original remit."

CFA UK – Investment professional body (#CL111)	"There is a growing need for guidance on [ETS], especially once allowances are auctioned. Standards on intangibles and inventory seem relevant, so the priority should be to see whether the measurement issue can be solved by applying existing standards."	EU Round Table of Industrialists (#CL127)	"[W]e consider that existing guidance in IAS20 and IAS37 is sufficient to address this issue which should be removed from the agenda."
Ernst and Young Global Ltd (#CL248)	"This is currently a gap in IFRS that is leading to significant diversity in practice and different accounting results . . . Emission reduction schemes continue to evolve and develop around the globe that means this project will have a more widespread impact on IFRS . . . If undertaken, this would be a comprehensive project requiring significant effort by the IASB that also considers the interaction with accounting for government grants and intangibles."	International Energy Accounting Forum (IEAF) (#CL122)	"The IEAF does not believe that it would be necessary for the IASB to give priority to [this] project. . . . As it is a key topic for our industry we have followed the discussions of the IASB about this topic with great interest. In our opinion, it is of high importance that the accounting model for emission rights fairly presents the economic substance of the business models of the companies. In particular, 'artificial' income volatility that is only accounting driven, but does not exist in economic reality because the entity is completely hedged, has to be avoided."
The Hundred Group of Finance Directors (#CL227)	"We recognise that ETSs are increasing in importance as more and more countries seek to meet emissions targets. We therefore believe that the Board should continue with this project. We are mindful that this is likely to be a difficult project . . . and recognise that it is likely to require considerable resources."	Australian Accounting Standards Board (AASB) (#CL237)	"[W]hilst the need for a specific project. . . seems initially attractive, the AASB believes it highlights that existing IFRSs are deficient in the sense that they are too narrowly focused. If IFRSs . . . were sufficiently broad in scope and sufficiently principle-based, accounting for new topics such as ETS could be determined within the context of those standards. Writing rule-based solutions for each new topic as they arise will add to the complexity of IFRSs and the inconsistencies between them."

The case study illuminates how in practice markets are not, of course, abstract entities, but rather networks of things, people and institutions that are particular to different places, times and commodities. This local embeddedness of markets poses a challenge for standards and other acts of commensuration, which attempt to harmonise and make things the same across space. This is especially true for financial accounting, where the IASB aims to provide an international set of standards underpinning corporate activity worldwide. By investigating a particular empirical case that has remained stubbornly outside of international financial accounting standards – that of the greenhouse gas emission allowance – and where local and regional standards have begun to emerge, it directs our attention to questions of market stability, and what happens in cases of prolonged non-standardisation.

Part of the problem with carbon financial accounting (and the reason for an extended delay in coming to a resolution) is the IASB objective of a 'neat' solution – i.e. one that complies with existing standards and leaves them unchanged. Carbon financial accounting national and sectoral standard setters have been able to accept a more practical approach and have 'muddled through', in order to adapt and prepare guidance more readily than international standard setters. Working 'at the coalface', the national and sector standard setters concerns have naturally been more centred on market stability than the protection of a set of principles-based, conceptual accounting rules (which nonetheless are central to the identity of the IASB). So this is an empirical case study about the malleability (or not) in practice of existing standards. Further, the geography of these standards is of interest: the IASB with its international scope is held in high esteem, but as national and regional standards gain traction it is as yet unclear how these different geographies of emission allowances standard settings will play out.

It is in the detailed political economy of carbon markets that key decisions have been made, but with not all types of expertise at the table. The conception of carbon as a commodity to be traded in markets was not something that financial accounting institutions were part of, and yet nevertheless the carbon commodity has crossed institutional borderlands into financial accounting, in the process revealing much about the assumptions, expertise and knowledge embedded within financial accounting standards, as well as the limitations of carbon markets. Technical expertise is split across the IASB (financial accounting) and the UNFCCC and European Commission (carbon markets), and these policy and standard setting organisations – split as they are into distinct roles – have yet to grasp the breadth and depth of the implications of an absence of international standards in the way that those 'in the wild' – companies, auditors, etc. – whose work does intersect (in a day-to-day, practical way) financial accounting and carbon markets do comprehend. As noted, it is these organisations that have therefore been the ones developing guidance and standards.

A case has been made here for greater consideration of the acts of commensuration integral to making markets and economies – in particular, to the work standards do to harmonise practices and activities, and how climate change, in manifold ways, gives rise to new things and issues that challenge existing standards. Often discounted as mundane, and typically obscured by technical language, standards such as those in financial accounting are nevertheless essential to understanding the day-to-day operation of carbon markets, as well as the implementation of other climate change policies. Climate change is a relatively new problem, and it causes friction and difficulties because existing standards are of limited utility – climate change was not a problem in the past – and these existing standards may therefore hinder rather than help processes of policy and technology change. The case of carbon financial accounting reveals how making climate change fit is a complex, uncertain and messy sociotechnical process, offering valuable insights and significant scope for additional empirical research across other standard setting arenas. In particular, it demonstrates the possibility of a wider conceptualisation of the complex geography of international standards, considering not just the hidden nature and invisibility of these standards, but also the emergence of regional or local standards (perhaps in competition with one another) and the implications of non-existent standards.

Bibliography

Abbatte, J. (1999). "From control to coordination: new governance models for information networks and other large technical systems". *The Governance of Large Technical Systems London & New York*. O. Coutard (ed.). London and New York, NY, Routledge: 114–129.

Ascui, F. and H. Lovell (2011). "As frames collide: making sense of carbon accounting". *Accounting, Auditing and Accountability Journal* **24**(8): 978–999.

Australian Accounting Standards Board (AASB) (2012). "Issues paper: approaches to recognition and measurement of emission liabilities arising from Emissions Trading Schemes". Melbourne, Australian Accounting Standards Board.

Autorite des Normes Comptables (ANC) (2012). "Proposals for accounting of GHG emission rights". Paris, Autorite des Normes Comptables (ANC).

Bailey, I. and S. Maresh (2009). "Scales and networks of neoliberal climate governance: the regulatory and territorial logics of European Union emissions trading". *Transactions of the Institute of British Geographers* **34**(4): 445–461.

Bailey, I. and G. A. Wilson (2009). "Theorising transitional pathways in response to climate change: technocentrism, ecocentrism and the carbon economy". *Environment and Planning A* **41**: 2324–2341.

Bebbington, J. and C. Larrinaga-Gonzalez (2008). "Carbon trading: accounting and reporting issues". *European Accounting Review* **17**(4): 697–717.

Bowker, G. C. and S. L. Star (2000). *Sorting Things Out: Classification and its Consequences*. Cambridge, MA, The MIT Press.

Callon, M. (1998). "An essay on framing and overflowing: economic externalities revisted by sociology". *The Laws of the Markets*. M. Callon (ed.). Oxford, Blackwell: 244–269.

Callon, M. (2007). "What does it mean to say that economics is performative?" *Do Economists Make Markets? On the Performativity of Economics*. D. MacKenzie, F. Muniesa and L. Siu (eds). Princeton, NJ, Princeton University Press: 311–357.

Callon, M. (2009). "Civilizing markets: carbon trading between *in vitro* and *in vivo* experiments". *Accounting, Organizations and Society* **34**(3–4): 535–548.

Callon, M. and F. Muniesa (2007). Economic experiments and the construction of markets. *Do Economists Make Markets? On the Performativity of Economics*. D. MacKenzie, F. Muniesa and L. Siu (eds). Princeton, NJ, Princeton University Press: 163–189.

Cook, A. (2009). "Emission rights: from costless activity to market operations". *Accounting, Organizations and Society* **34**(3–4): 456–468.

Cronon, W. (1991). *Nature's Metropolis: Chicago and the Great West*. New York, NY, Norton and Company.

Deloitte. (2012). "Accounting for emission rights an urgent topic for national standard setters". Online at: www.iasplus.com/en/news/2012/may/accounting-for-emission-rights-an-urgent-topic-for-national-standard-setters (accessed 25 May 2012).

Edwards, P. N. (2004). "'A vast machine': standards as social technology". *Science* **304**(5672): 827–828.

Espeland, W. N. and M. L. Stevens (1998). "Commensuration as a social process". *Annual Review of Sociology* **24**: 313–343.

Grubb, M. J. (2002). "Britannia waives the rules: the United Kingdom, the European Union and climate change". *New Economy* **9**(3): 139–142.

Hatherly, D., D. Leung and D. MacKenzie (2008). "The finitist accountant". *Living in a Material World: Economic Sociology Meets Science and Technology Studies*. T. Pinch and R. Swedberg (eds). Boston, MA, MIT Press: 131–160.

Higgins, V. and W. Larner (2010). *Calculating the Social: Standards and the Reconfiguration of Governing*. London, Palgrave Macmillan.

Hopwood, A. G. and P. Miller (eds) (1994). *Accounting as Social and Institutional Practice*. Cambridge Series in Management. Cambridge, Cambridge University Press.

International Accounting Standards Board (IASB) (2008). "International Accounting Standards Board (IASB) – information for observers: Emissions Trading Schemes; board meeting 20 May 2008". Online at: www.iasb.org/NR/rdonlyres/92B01EDC-E519–431F-915F-0F33505D7DFD/0/ETS0805b03obs.pdf (accessed 3 October 2008).

International Accounting Standards Board (IASB) (2009). "Emissions Trading Schemes: accounting for the right to future installments. Agenda paper 18". Online at: www.ifrs.org/Current-Projects/IASB-Projects/Emission-Trading-Schemes/Meeting-Summaries/Documents/ETS1209b18obs.pdf (accessed 30 August 2014).

International Accounting Standards Board (IASB) (2012a). "Feedback statement: Agenda Consultation 2011". London, International Accounting Standards Board (IASB).

International Accounting Standards Board (IASB) (2012b). "IASB – who we are and what we do". Online at: www.ifrs.org/About-us/Pages/Who-We-Are.aspx (accessed 21 August 2014).

International Accounting Standards Board (IASB) (2012c). "IFRS staff paper for IASB meeting May 2012: Agenda Consultation 13B". London, International Accounting Standards Board (IASB).

International Energy Accounting Forum (IEAF) (2010). "International Energy Accounting Forum IFRS paper – accounting for emission allowances". Geneva, International Energy Accounting Forum.

International Financial Reporting Standards (IFRS) (2012). "IASB/FASB meeting May 2012". Online at: www.ifrs.org/Meetings/IASB+May+2012.htm (accessed 30 May 2012).

Kolk, A., D. Levy and J. Pinkse (2008). "Corporate responses in an emerging climate regime: the institutionalization and commensuration of carbon disclosure". *European Accounting Review* **17**(4): 719–745.

KPMG (2008). "Accounting for carbon: the impact of carbon trading on financial statements". London, KPMG.

Lohmann, L. (2009). "Toward a different debate in environmental accounting: the cases of carbon and cost-benefit". *Accounting, Organizations and Society* **34**(3): 499–534.

Lovell, H. and D. MacKenzie (2011). "Accounting for carbon: the role of accounting professional organisations in governing climate change". *Antipode* **43**(3): 704–730.

Lovell, H., J. Bebbington, C. Larrinaga and T. Sales de Aguiar (2013). "Putting carbon markets into practice: a case study of financial accounting in Europe". *Environment and Planning C* **31**: 741–757.

Lovell, H., T. Sales de Aguiar, J. Bebbington and C. Larringa-Gonzalez (2010). "Accounting for carbon – ACCA & IETA research report 122". London, The Association of Chartered Certified Accountants (ACCA) and the International Emissions Trading Association (IETA).

MacKenzie, D. (2008). "Producing accounts: finitism, technology and rule following". *Knowledge as Social Order: Rethinking the Social Theory of Barry Barnes*. M. Mazzotti (ed.). Aldershot, Hants, Ashgate: 99–117.

MacKenzie, D. (2009). "Making things the same: gases, emission rights and the politics of carbon markets". *Accounting, Organizations and Society* **34**(3–4): 440–455.

Muniesa, F., Y. Millo and M. Callon (2007). "An introduction to market devices". *Market Devices*. M. Callon, Y. Millo and F. Muniesa (eds). Oxford, Blackwell Publishing/The Sociological Review: 1–12.

Point Carbon (2007). "Carbon 2007: a new climate for carbon trading". Online at: www.pointcarbon.com/research/carbonmarketresearch/analyst/1.189 (accessed 21 August 2014).

Tanaka, K. and L. Busch (2003). "Standardization as a means for globalizing a commodity: the case of rapeseed in China". *Rural Sociology* **68**(1): 25–45.

Timmermans, S. and S. Epstein (2010). "A world of standards but not a standard world: toward a sociology of standards and standardization". *Annual Review of Sociology* **36**: 69–89.

8

SUMMARY AND CONCLUSIONS

The Making of Low Carbon Economies analyses how sustained effort at climate change mitigation over two decades has resulted in a variety of new institutions, ideas, rules and practices. The last twenty years have seen active construction of low carbon economies. The basic premise of this book is that, given that low carbon economies are a mix of things and people, responses to climate change can be best understood by looking in detail at the heterogeneous networks where the reality of climate change is recognised through its incorporation into everyday lives. 'Everyday lives' in this context refers mostly to the day-to-day lives of professionals, working across a range of sectors, rather than the more typical focus on the domestic realm. This book has documented an ongoing process of stitching climate concerns into the discourse and practices of already existing economies and markets, as well as the process of creating discrete new markets specifically aimed at climate mitigation.

Climate change is predominately framed as an economic problem, as something to be solved through economies and markets. This neoliberal approach is often criticised, but if one adopts a broader understanding of markets and economies, following the lead of Foucault (2007) and other scholars, such as Mitchell (2008, 2011) and Gibson-Graham (2006, 2008), then the making of low carbon economies can be rethought more positively. These economies can be seen as inclusive and open to being remade according to a wider range of ideas, objectives and practices and not necessarily limited by a capitalist focus on profits and monetary exchange. The empirical chapters also indicate how our response to the problem of climate change typifies the blurred boundaries that exist in practice between markets and economies (Lie 1997; Fligstein and Dauter 2007), with such issues as value, commodification and standardisation transcending discrete low carbon markets and 'overflowing' (Callon 1998) into the political and governance arenas embraced by broader economies.

Instead of concentrating on the discipline of economics, *The Making of Low Carbon Economies* attempts an interdisciplinary synthesis of concepts and theories drawn from sociology, human geography, STS and Foucauldian scholarship.

There are two main reasons for this interdisciplinary approach. First, a desire to encourage a focus on micro-scale economic processes; the everyday, often banal, practices and ways of doing that together constitute particular markets and economies. This approach necessarily includes attention to people, discourse, materials, institutions, values, policy networks and relations: diverse elements that are hard to capture using just one disciplinary lens. Second, the observation that climate change is being understood in relation to existing ways of doing, expertise and knowledge and attempts to mitigate climate change can therefore only be fully comprehended in relation to what has gone before. This book thus draws on transitions theory to consider these processes and patterns of change over time, including the catalysts of innovation (Geels 2011; Markard *et al.* 2012), as well as Foucault's archaeological approach to researching the changing meaning of key terms, disciplines and bodies of knowledge over time (Rabinow 1984; Tilley 1990; Foucault 2007).

The empirical chapters of *The Making of Low Carbon Economies* reveal that a significant shift in our capabilities and responses to climate change is underway. This shift is evident when we examine the development of low carbon economies and markets across many different sites, at a micro-scale. If we 'lift the stone' on such topics as house building in the UK, provision of energy in cities and financial accounting, we observe how carbon as a commodity, and climate change as an issue, are gradually being woven, often in quite subtle ways, into the daily operation and decision-making of markets and economies. The findings reported in this book provide grounds for hope for those of us who wish to act in response to the problem of climate change, and thereby some guidance for those who make policy. Overall, the book presents a more optimistic picture of society's response to climate change than is found within 'top-down', large-scale international reviews of climate mitigation policy and activities (see, for example, UNFCCC 2012, 2014; Stocker *et al.* 2013). Here, I provide evidence of the many ways in which climate change is being incorporated and taken seriously within existing practices, standards and professions, across several different empirical sites. Of course, these changes are not without tension, but this book demonstrates how a number of important low carbon transitions are already well underway. These changes range from the growing attention of the International Accounting Standards Board (IASB) to carbon (Chapter 7), to the work by remote sensing scientists to improve forest carbon measurement techniques (Chapter 6), and the conjunction of climate change, fuel poverty and housing modernisation objectives resulting in the implementation of district heating in social housing (Chapter 3). These achievements do not make headline news, because they are in many cases both technical and mundane and take place mostly away from public arenas. But these largely hidden advances demonstrate that many things are, in fact, happening to help solve the problem of climate change.

The three themes that structure this book – heterogeneous networks, framing and commensuration – are designed to focus attention on this underlying detail of low carbon markets and economies. The Part I theme of heterogeneous networks concentrates our gaze on the mix of things and people that comprise markets and economies. Framing, the Part II theme, reveals the value of creating boundaries around ideas, issues, ways of doing and expertise, but also demonstrates the limitations of frames, which can restrict novelty, innovation and learning. The Part III theme of commensuration concentrates on the intricate and usually hidden work of calculation, standardisation and measurement that forms the calculative bedrock of markets and economies. The overlaps between the three themes are evident. In Chapter 4, for example, we see how frames are created and sustained, not just through discourse, but also by material things – in this case, low energy housing. Similarly, in Chapter 6, it is described how the pervasive discourse of 'measurement, reporting and verification' in forest carbon markets has emerged, and is sustained, through certain types of calculative expertise – in particular, by remote sensing.

The three themes, used in conjunction, help us understand why there is no clear disjuncture with the arrival of climate change as a problem and its acceptance, but rather a gradual, messy process of weaving climate change into existing markets and economies and ways of doing. As noted, looking for big 'headline news' about solutions to climate change can be unhelpful. Significant progress in mitigating climate change is instead to be found in a quite different form, across a range of geographical and professional sites and incorporated into mundane daily practices and objects. A common theme running through the three parts is the obscurity or 'hiddenness' of much of the work involved in the making of low carbon economies. Discrete carbon markets, purposefully made to mitigate climate change, as explored in Chapter 2, are therefore atypical. Carbon markets are to date the most visible face of the making of low carbon economies, but here I attempt to show that the remaking of already existing economies and markets is just as interesting as the newly created carbon markets, and arguably more significant.

Use of the three themes of heterogeneous networks, framing and commensuration helps to capture the sheer diversity of approaches by which low carbon economies are being made, as is demonstrated in case studies of housing, energy, forests and accounting. *The Making of Low Carbon Economies* reveals how low carbon economies are variously constructed and shaped through such processes as (re)framing, standardisation and calculation. Moreover, there is evidence of the growing stability of these efforts – for example, through their development into new fields of expertise (such as carbon accounting; see Chapter 5) and incorporation into standards and policies (such as the 2016 UK zero carbon housing policy discussed in Chapter 4, and the plethora of new forest carbon monitoring standards documented in Chapter 6).

There is strength in diversity: the analysis within this book shows that there are many potential avenues for mitigating climate change. This finding aligns with current scholarship on 'diverse economies', which considers the multiple ways of being economic and encourages attention to this diversity (Smith 2004; Leyshon 2005; Gibson-Graham 2008). It also fits closely with the notion of sociotechnical transitions, which likewise embraces diversity in innovation, at least at the level of the 'innovation niche'. Transitions theory captures the constant tension between continuity and change identified in this book. There is a constant 'bubbling up' of innovations at a niche level, and some of these innovations eventually have a significant influence on dominant economies and markets – i.e. on the mainstream sociotechnical regime. However, in the case of the climate change mitigation case studies identified and explored here, there is a more diffuse and erratic boundary between new ideas and established ways of doing, between niches and regimes. A mishmash of different, new low carbon ideas, technologies and practices is becoming incorporated in a rather haphazard and piecemeal way into existing markets and economies. Chapter 2, for example, describes a 'pick and mix' approach to learning from the voluntary offset market innovation niche, with some ideas and technologies taken up within the compliance offset markets, but others not. In Chapter 4, it is similarly observed how ideas and technologies from the low carbon housing 'innovation niche' that emerged in the UK at the turn of the century, and were then 'mainstreamed' by the UK Government with their radical 2016 zero carbon homes policy, have been demonstrated as problematic and not truly tested at the niche stage. These limitations associated with the overly neat categories of transitions theory, and its temporal thrust, have been noted elsewhere (Hodson and Marvin 2010; Coenen and Truffer 2012; Smith and Raven 2012; Bridge *et al.* 2013).

The Making of Low Carbon Economies demonstrates the breadth of these low carbon economies, their geographies and histories (Smith 2004; Oberhauser 2005; Gibson-Graham 2008). Low carbon markets and economies are made in specific places, in localities as well as in arenas of professional expertise, with particular ways of thinking and doing. How low carbon economies are made depends on these different contexts. Attention to heterogeneous networks allows us to embrace the particularity of any one low carbon economy or market. By also incorporating the themes of framing and commensuration, we can understand how new, initially context-specific, climate change ideas, practices and technologies might diffuse and become more widely accepted and used. Frames provide a basis for making sense of new problems and their solutions and hence are crucial to the diffusion of new low carbon ideas and techniques. Acts of commensuration represent the more practical 'flipside' of frames, by allowing new discourses to come into being through their translation into modes of calculation and standards. These modes of calculation and standards in turn influence frames; there is a two-way interplay. Such an approach maps reasonably well onto transitions theory, with processes of framing, standardisation and measurement most pertinent to sociotechnical regimes; whilst a heterogeneous networks perspective has synergies with innovation niches, the small-scale sites where radical innovations first develop through forging new sociotechnical relations.

A further valuable outcome of using an interdisciplinary approach is in directing our attention to the particular histories of the making of low carbon economies. I have presented evidence in a number of chapters of the role of prior types of expertise and material infrastructures in shaping low carbon economies, thereby demonstrating their origins. Chapter 7, for instance, explores how carbon has been understood, framed and classified by accountants with reference to other, more familiar things, such as government grants, taxes and leases. The study of low carbon economies is hence not just about the development of new climate change related standards, practices and knowledge, but also about the reshaping of existing practices and bodies of expertise, which in many cases are extensive and have been developed over many decades. Nonetheless, there is evidence that they are changing and evolving in response to climate change. Even small shifts, for example, in the mostly hidden acts of commensuration underpinning markets and economies, designed to make them attentive to the problem of climate change, will collectively have significant ramifications.

This book is aimed at an academic audience, rather than at those who make policy. Yet the evidence presented here should give pause for thought to those who are ideologically opposed to mitigating climate change through markets and economies. I hope to have demonstrated that markets and economies have always been made – discursively, materially, cognitively – and therefore that they can be remade differently, including through incorporating climate mitigation. Further, diverse economies have always existed side-by-side, and will continue to do so (Leyshon 2005; Gibson-Graham 2008). There are hence many different ways of responding to the problem of climate change, and what this book captures and conveys is this potential for diversity. Much of the already existing progress towards finding solutions to the problem of climate change is, however, obscured by a focus on large, 'top-down' linear responses to climate change. As noted, looking for big 'headline news' about solutions to climate change presents an overly negative picture of the significant headway that has already been made. Evidence of the making of low carbon economies is instead to be found in a quite different form, across a range of geographical and professional sites and incorporated into mundane daily practices, discourse, standards and artefacts.

Bibliography

Bridge, G., S. Bouzarovski, M. Bradshaw and N. Eyre (2013). "Geographies of energy transition: space, place and the low-carbon economy". *Energy Policy* **53**: 331–340.

Callon, M. (1998). "An essay on framing and overflowing: economic externalities revisted by sociology". *The Laws of the Markets*. M. Callon (ed.). Oxford, Blackwell: 244–269.

Coenen, L. and B. Truffer (2012). "Places and spaces of sustainability transitions: geographical contributions to an emerging research and policy field". *European Planning Studies* **20**(3): 367–374.

Fligstein, N. and L. Dauter (2007). "The sociology of markets". *Annual Review of Sociology* **33**: 105–128.

Foucault, M. (2007). *Security, Territory, Population*. Basingstoke, Palgrave Macmillian.

Geels, F. W. (2011). "The multi-level perspective on sustainability transitions: responses to seven criticisms". *Environmental Innovation and Societal Transitions* **1**(1): 24–40.

Gibson-Graham, J.-K. (2006). *A Post-Capitalist Politics*. Minneapolis, MN, University of Minnesota Press.

Gibson-Graham, J.-K. (2008). "Diverse economies: performative practices for other worlds". *Progress in Human Geography* **32**(5): 613–632.

Hodson, M. and S. Marvin (2010). "Can cities shape socio-technical transitions and how would we know if they were?" *Research Policy* **39**: 477–485.

Leyshon, A. (2005). "Introduction: diverse economies". *Antipode* **37**(5): 856–862.

Lie, J. (1997). "Sociology of markets". *Annual Review of Sociology* **23**: 341–360.

Markard, J., R. Raven and B. Truffer (2012). "Sustainability transitions: an emerging field of research and its prospects". *Research Policy* **41**(6): 955–967.

Mitchell, T. (2008). "Rethinking economy". *Geoforum* **39**: 1116–1121.

Mitchell, T. (2011). *Carbon Democracy: Political Power in the Age of Oil*. London, Verso Books.

Oberhauser, A. M. (2005). "Scaling gender and diverse economies: perspectives from Appalachia and South Africa". *Antipode* **37**(5): 863–874.

Rabinow, P. (ed.) (1984). *The Foucault Reader: An Introduction to Foucault's Thought*. St Ives, Clays Ltd.

Smith, A. (2004). "Regions, spaces of economic practice and diverse economies in the 'new Europe'". *European Urban and Regional Studies* **11**(1): 9–25.

Smith, A. and R. Raven (2012). "What is protective space? Reconsidering niches in transitions to sustainability". *Research Policy* **41**(6): 1025–1036.

Stocker, T. F., D. Qin, G.-K. Plattner, M. Tignor, S. K. Allen, J. Boschung, Y. Nauels, V. Xia and B. A. P. M. Midgley (eds) (2013). *Climate Change 2013: The Physical Science Basis. Contribution of Working Group I to the Fifth Assessment Report of the Intergovernmental Panel on Climate Change*. Cambridge and New York, NY, Cambridge University Press.

Tilley, C. (1990). "Michel Foucault: towards an archaeology of archaeology". *Reading Material Culture: Structuralism, Hermeneutics and Post-Structuralism*. C. Tilley (ed.). Oxford, Basil Blackwell: 281–347.

United Nations Framework Convention on Climate Change (UNFCCC) (2012). "The CDM policy dialogue: background". Online at: http://cdm.unfccc.int/about/policy/index.html (accessed 19 March 2014).

United Nations Framework Convention on Climate Change (UNFCCC) (2014). "Press release – negotiations towards new universal climate agreement in 2015 get underway". Online at: https://unfccc.int/files/press/press_releases_advisories/application/pdf/20141403_adpclose.pdf (accessed 17 March 2014).

INDEX

Page numbers in *italics* denote tables, those in **bold** denote figures.

actor-networks/actor network theory 17–18
advocacy coalition, definition 84
agencement 2, 14, 17–20, 22
agency 15, 17–18, 76–7
Ahuja, D.R. 110
aims and objectives of the book 1–3
ANC (Autorité des Normes Comptables) 163
Annex One countries 137
Arrhenius, S. 107–8
Ascui, F. 119n1, 106
assemblage 14, 17–18
audience, target 3–4
Avoided Deforestation 43–4; *see also* REDD+

Bailey, I. 160
Baker, J. 141
Bali Action Plan 136
Bali COP 137
Barry, A. 14, 126
BedZed (Beddington Zero Energy Development) 82, 89, 94
biomass technologies, for carbon offsetting 33
black boxing 17, 38
boundary organisations, role of 110
Boushel, C. 38
boutique carbon 35, 40
Bowker, G.C. 126, 130
Boyd, E. 136
"Building a greener future" (DCLG 2006) 95
"Building a low-carbon economy: the UK's contribution to tackling climate change" (UK Climate Change Committee) 4
Bumpus, A. 38
bureaucracy, impact of 44

Çalişkan, K. 5, 76, 129
Callon, M. 5, 17–21, 72, 76–7, 117, 129
capitalism, disciplinary technologies as precursor to 16
carbon accounting: defining 104–5, 117–19; diversity of practices 113; financial framing 113–14 (*see also* financial accounting); forest carbon markets *see* forest carbon MRV; and GWP 110–11; key findings 104; market-enabling framing 111–13; multiple frames of 107; physical framing 107–8; political framing 109–11; and REDD+ 113 (*see also* REDD+); role of boundary organisations 110; social/environmental framing 114–17; storage (sinks) 112–13; summary and conclusions 119; UNFCCC requirements 109; UNFCCC– Kyoto Protocol discrepancies 111–12
carbon accounting guidance, KPMG 163
carbon markets 30–6
carbon offsetting: air travel offset requirements, divergence in estimates

108; 'boutique' or 'gourmet' carbon 35, 40; compliance markets vs voluntary 31–2, 35, 42–5; compliance offset market, objective 27; European Emissions Trading Scheme 31, 113–14, 116–17, 159–61, 164, 167; fungibility 35, 38; heterogeneous networks, achieving credibility in 36–41; innovation niche, the voluntary offset market as 38, 42–5; key findings 27; market creation implications 31; market share of different technologies 32; and the migration of techniques 39; MRV, role of 36, 38; and the path dependency of technologies 41–2; price of the CER 35; sociotechnical transitions and carbon offset markets 41–2; standards and regulation 31–2; summary and conclusions 46–8; technologies, diversity of 33; technologies used in the CDM *33*, *34*; technology, theorising the role of in offsetting 36; the voluntary offset market 27, 31 (*see also* voluntary offset market)
carbon sinks, examples 112
carbon trading: impact on corporate behaviour 116; *see also* emissions trading
CAT (Centre for Alternative Technology) 84, 86, 94
CDM (Clean Development Mechanism): calculation, obsession with 36, 38; carbon offsetting technologies *33*, *34*; and the creation of emission rights 31; criticisms 44; and deforestation reduction 113, 137; emissions reduction targets 30–1; function 20, 30; governance process 37; governance process, criticisms of the 37; "making things the same" 112; offset projects, most common types 32; Policy Dialogue initiative 46; risk-averse approach 43; technology approval process 32;
CDP (Carbon Disclosure Project) 115–16
CERs (Certified Emission Reductions) 31, 35, 112
CESP (Community Energy Saving Programme) 58–9
Climate Disclosure Standards Board 117
climate mitigation: and the 'measure to manage' rationality 135; optimism about 29
'climategate' leaked emails furore 108
CO_2 reduction targets, UK 83; *see also* zero carbon 2016 policy
Collins, H.M. 15
commensuration: the concept of 125–30; definition 3; and forest carbon MRV 140, 150–1 (*see also* forest carbon MRV); and governmentality 127, 133; importance for climate change 3, 126; key insights 125; sociotechnical nature 128; and standard setting 127–9
"common but differentiated responsibilities" 109, 112
compliance carbon markets, innovation constraints 38
compliance carbon offsets, achieving fungibility 35
conclusions, summary and 174–8
cookstoves 40–1
corporate sustainability reporting 115
cost-effectiveness, as evaluation criteria for low carbon housing 92

Dean, M. 4
deep green values 84–5, 91
Deforestation, Avoided 43–4; *see also* REDD+
deforestation, tropical, estimated GHG production 134
Deleuze, G. 14
Deloitte 163
disciplinary technologies: examples of 16; as precursor to capitalism 16
discourse coalitions, definition 75
disentanglement 76, 129
district heating, residential *see* residential district heating
Dunn, E.C. 126

ecological footprints 116
ecological modernism, definition 90–1
economies, vs economy 6–7, 178
economy: the concept of 5–6; Foucault on the evolution of the meaning 6
Edwards, P.N. 128

emission allowances 20, 105, 130, 156–67, 170; roots of standard-setting challenges 164
emissions trading: European scheme (EU ETS) 113–14, 116–17, 159–61, 164, 167; extracts of the IASB 2011 Agenda Consultation response letters related to *168–9*; IASB-FASB joint project 161; schemes 105, 165–6
energy efficiency 4, 57, 59, 89, 92
Epstein, S. 126, 128
Espeland, W.N. 130
Etzion, D. 119

Ferraro, F. 119
financial accounting: commensuration scholarship related to 158; comparison of carbon to other commodities 165–6; disclosure reduction 166–7; extracts of the IASB 2011 Agenda Consultation response letters related to emissions trading *168–9*; IASB lobbying 166; 'incommensurables' examples 160; key findings 156; research material 157; responses of accountants to the absence of international carbon accounting standards 162–7; role of auditors 163; seeking authority beyond IASB 162–5; standards and emission allowances 158; standards setting 158–9; summary and conclusions 167–71
financial carbon accounting, global standards 114
Findhorn Ecovillage, Scotland 84
flexibility mechanisms 111
Fligstein, N. 5
forest carbon MRV: baseline standard 147; central role of MRV within forest carbon markets 138–9; challenges 150–1; consensus, governmentality perspective on 139–40; defining practices and technologies 135; dominant discourse 140–1; examples of use *142–3*; excavating tree roots for carbon measurement **134**; forest carbon markets, brief introduction 136–8; ground-based local community forest monitoring 150; impact of excessive focus on 133; IPCC guidance 138–9; key findings 133; levels of accuracy 147, 150; need to develop 'compliance-grade' systems 140; optimism about 140–1; participating organisations 141; remote sensing, importance of and limitations 141; remote sensing techniques 144–7; socio-technical perspective 150; standards, examples of *148–9*; standards assessment 147–51; summary and conclusions 151–2; understanding the vast amount of work required 140; UNFCCC requirements 150; *see also* REDD+
Foucault, M. 6, 16, 73–5, 135
fragility, of heterogeneous networks 52, 67, 76 (*see also* residential district heating)
frame-reflective discourse 73, 119
framing: and the claiming of ownership 75; the concept of 71–8; concept origins 2; in discourse and practice 72–3; as essential precursor to action 72; fragility of 76; and governmentality 72; key insights 71; lock-in and 21; origins and development 71–2; and significant change 77
fuel poverty 52, 58–9, 62, 89–90

GBC (Green Building Council) 97
GHG Protocol 115
GHGs: emissions rights and obligations 113; impact evaluation tools 110; reduction targets, UK 83; tropical deforestation, estimated percentage from 134
Gibson-Graham, J.-K. 7
Global Canopy Programme 138
global carbon cycle, first quantitative account of the 107
Global South, and the voluntary offset market 32, 43
global warming, scientific concern about 108
gourmet carbon 35, 40
governmentality: commensuration and 127, 133; and forest carbon MRV 135, 139–40; Foucault's concept of 16, 135, 139; framing and 72

Gray, R. 114
greenhouse gases, definition 48n2 (*see also* GHGs)
GRI (Global Reporting Initiative) 115–16
Grubb, M.J. 160
Guattari, F. 14
GWP (Global Warming Potential): development of the framework 110; and the economic framing of climate change 111; Kyoto mandate 110

Hajer, M. 72, 75, 91
Hardie, I. 18
Harvey, D. 74
heterogeneous networks: and actor-network theory 17–18; and *agencement* 18, 22; the concept of 13–23; concept origins 14; and the creation of markets and economies 19–20; credibility, achieving with carbon accounting 36–41; fragility of 52, 67, 76 (*see also* residential district heating); and governmentality 16; key insights 13–14; materiality 15; voluntary offset markets as 39–40
Hewitt, Patricia 82
Higgins, V. 126
Hockerton Housing Development, Nottinghamshire 82
Hogbom, A. 107–8
Honduras, carbon offset projects 38
Hughes, T.P. 21, 23

IASB (International Accounting Standards Board): accounting guidance issued by 159; carbon classification and standardisation proposals 160–1; counterpart 157; emissions trading, consultation response letters *168–9*; establishment and location 158; and the global financial crisis 166; initial attempts to establish financial carbon accounting guidelines 113; membership 157; stakeholder lobbying 166; view of the 'activity-based' model 164
IEAF (International Energy Accounting Forum) 163
IFRIC (International Financial Reporting Interpretations Committee) 159
IFRIC-3 159–61
incommensurables 125, 130, 157, 160
Indonesia, Li's analysis of development projects in 37
innovation, the voluntary offset market as a niche for 38, 42–5
institutional ambiguity, in the environmental debate 72
interdisciplinary themes of the book 1–2
IPCC (Intergovernmental Panel on Climate Change): establishment and function 109; Guidelines, publication 109; role of 110

Jasanoff, S. 4
Jenkins Smith, H.C. 81

Kern, F. 78
key terms of the book 4–7
Kolk, A. 115–16
KPMG 163
Kyoto Protocol: carbon accounting discrepancies with UNFCCC 111; exclusion of reduced deforestation from carbon markets 113; flexible mechanisms 30, 111; GHG emissions rights and obligations 113; GHG reduction targets, UK 83; GWP 110; "making things the same" mandate 112; *see also* CDM

La Esperanza project 38
Larner, W. 126
Lashof, D.A. 110
Law, J. 14, 40
LCA (life cycle analysis) 116–17
Li, T.M. 4, 37, 39, 126
Lie, J. 6
life cycle approach, to low carbon housing 91–3
Liverman, D. 29–30
lock-in 16–17, 20–1, 23
Lovbrand, E. 140
low carbon economy: the concept of 4; definitions 4–5
low carbon housing: definition 83;

low carbon housing *continued*
discursive storylines 91–5; financial investment discourse 92; framing sustainable housing as a solution 86–91; key findings 81; life cycle approach 91–3; and payback 92–3; smart housing approach 93–5; summary and conclusions 98–9; sustainable housing advocacy coalition 84–6; transitions theory perspective 95–8; UK zero carbon 2016 policy 82, 95–8; *see also* residential district heating
low carbon technologies, reliability as market actors 52

MacKenzie, D. 4, 14, 18, 135
The Making of Low Carbon Economies: aims and objectives 1–3; interdisciplinary themes 1–2; key terms 4–7; structure 7–10; summary and conclusions 174–8; target audience 3–4
market devices, the notion of 15
markets: blurred boundaries between economies and 6; definitions 5–6
materiality 14–15, 19, 23, 38
Mathews, M.R. 115
Mauna Loa Observatory 108
McCarthy, J. 74
methane, GWP 111
Miller, C. 109
Milne, M. 111
Mitchell, T. 16, 21, 129
Montreal Protocol 109–10
Muniesa, F. 21
Munro, M. 14
Murdoch, J. 17

neoliberalism 74
Networks of Power (Hughes) 21
New Zealand 111
NGOs (non-governmental organisations), and the voluntary offset market 32

oil industry, Mitchell's analysis 16
overflows/overflowing 14, 21, 76–7, 117, 160–1

pacifying goods 76, 128, 160
performativity, in heterogeneous networks 19
photovoltaic panels, payback period 93
Plan Vivo 43
Plattner, G. 110
Prudham, S. 74

Rabinow, P. 4, 16
44
REDD+ (Reducing Emissions from Deforestation and Forest Degradation): basic idea 134; basis for measuring forest carbon under 135; a brief introduction 136–8; carbon measurement challenges 150–1; complexity of discussions around 139; core objective 136; creation of 128; influencing spheres 141; "Little REDD+ book" 138; measurement requirements 140; operationalisation debate 138; policy objective 134; potential benefits for non-Annex One parties 137; and the CDM 44
reduced deforestation, exclusion from carbon markets 113
Rein, M. 73, 77, 117
relational economic geography 19
remote sensing: competing measurement techniques 150; importance of and limitations 141; techniques for measuring forest carbon 144–7
residential district heating: and the 'bedroom tax' 66–7; billing problems 67; Council Liaison officer, appointment of 60; the Council Liaison Officer's activities and skills 62–3; expected cost to tenants 61–2; heating control panels **65**; heating control panels, complexity 64–6; heterogeneous network, fragility of in Fordside 59–62; housing developments, featured 54–6; installation decision-making process 57–9; key findings 52; objectives 59; pipework installation **57**; safety issues 57–8, 62; study location and chronology 54; summary and conclusions 67–8; tenant satisfaction 59; wiring damage 63–4

Rio Earth Summit 108
Ronan Point gas explosion 57–8, 62

Sabatier, P.A. 81
Schon, D. 73, 77, 117
SHQS (Scottish Housing Quality Standards) 57
Slater, D. 14
smart housing, definition 93
smart housing approach, to low carbon housing 93–5
Smith, A. 77–8
Smith, S.J. 14
social housing modernisation, and the making of new low carbon economies 58 (*see also* residential district heating)
Social Responsibility Accounting 115
sociotechnical transitions: and carbon offset markets 41–2; commensuration and 129; examples of 45; framing and 77–8; useful additions to a heterogeneous networks conceptualisation 23
sociotechnical transitions theory 22
standard setting, politics of 160–1
standards, ambiguity in making and operationalising 130
Star, S.L. 126, 130
Stern Report 4
Stevens, M.L. 130
Stripple, J. 140
structure of the book 7–10
summary and conclusions of the book 174–8
sustainable housing: definition 83; framing as a low carbon housing solution 86–91
sustainable housing advocacy coalition 86
sustainable housing developments: BedZed 82, 89, 94; examples 84; Hockerton 82

target audience of the book 3–4
technology, path dependency 41
terms, key, in this book 4–7
themes, interdisciplinary, in this book 1–2
Timmermans, S. 126, 128
transitions: definition 22; framing and 77–8; *see also* sociotechnical transitions
transitions theory: framing and 77; the 'multi-level perspective (MLP)' 22; perspective on low carbon housing 95–8; temporal thrust 23
tropical deforestation, estimated percentage of global GHGs 134
tropical forest ecosystems, fragility 151
Tyndall Centre 29–31, 35, 42, 44

UK: CO_2 reduction targets 83; UK zero carbon 2016 policy 82, 95–8
UNFCCC (United Nations Framework Convention on Climate Change): adoption 108; carbon accounting discrepancies between Kyoto Protocol and 111; carbon accounting requirements 109; carbon reporting requirements 137; and the compliance market 30–5, 39; and the role of forests in climate mitigation 137

Vale, B. and R. 93
Verified Carbon Standard 31
Versteeg, W. 72
voluntary offset market: advantages 45; bottom-up focus 39, 43; cookstoves, reframing of 41–2; core aim 40; development, independent 27, 31; flexibility 42–3; focus 35; informal nature 32; as innovation niche 38, 42–5; objective 27; standards 31

Williams, M. 133
Wilson, G.A. 160
World Bank 32
WWF (World Wide Fund for Nature) 88–9, 96–7

Yearley, S. 15

Zelizer, V. 5
zero carbon 2016 policy, UK government's 82, 95–8

For Product Safety Concerns and Information please contact our EU
representative GPSR@taylorandfrancis.com
Taylor & Francis Verlag GmbH, Kaufingerstraße 24, 80331 München, Germany

www.ingramcontent.com/pod-product-compliance
Ingram Content Group UK Ltd.
Pitfield, Milton Keynes, MK11 3LW, UK
UKHW020203200926
475702UK00004B/590
9780415724715